True courage isn't loud. It's the quiet, soul-wrenching decision to walk away from everything you once believed in, simply because truth demanded it. Mollie Engelhart did just that. In a world that clings to comfort and conformity, she chose principle over popularity, reality over ideology. Her journey is proof that bravery isn't about holding the line—it's about knowing when to draw a new one.

Alex Clark
Culture Apothecary podcast
Turning Point, USA

Humanity would be hopeless without the wisdom in *Debunked by Nature*! Our health is largely dependent on the health and stewardship of the soil that generates our food. Yet agendas, poisonous chemicals, and short-term profit motives try to outsmart nature's way. Similarly, the soil of our minds gets poisoned with digital reliance, agendas, and even well-meaning ideas and virtues laced with kindness that are dreadfully misaligned with our innate nature. As a regenerative farmer, mother, and truth seeker, Mollie brilliantly shines light on the parallels that run from our inner life, the external world, and the soil beneath our feet. She will help you wake up where you've been asleep and choose better where you've been fooled.

J.P. Sears
Comedian

It's hard to have courage in life as we grow, learn, and discover more than we probably anticipated. What I love about Mollie, and what she is doing, is the authentic nature of her approach and her curiosity with the world. She goes forth in truth, and I have so much respect for that.

Maggie Q
Actress, animal rights activist

Mollie's book comes at a crucial moment in Western culture. As postmodernism gives way to a kind of fluidity that rejects long-standing, tried-and-true wisdom, we find ourselves unmoored. In the vacuum, we've made countless attempts to reinvent the wheel when it comes to health, the environment, nutrition, family, and identity. Instead of acknowledging where we've gone wrong, we double down, embracing even more destructive trends, lifestyles, and ideologies. This book is a call to return to the place from which we fell. Whether she's confronting the natural cycle of life and death or exposing the myths that keep us medicated, dependent, and afraid, Mollie brings deep truths down to earth through the lens of her own story, with all of its highs, heartbreaks, and hard-earned wisdom. Her journey reminds us that the land we live on, the bodies we're given, the children we raise—all of it is on loan from God. And it's our responsibility to steward them well. Cheers to Mollie Engelhart, for reminding us that we are called to be faithful stewards. That we are not powerless to live well. We can grow food, raise families, seek the Lord, and honor both life and death without fear. That's not just possible; it's necessary.

Chad O. Jackson
Independent filmmaker

Mollie's story is nothing short of incredible. I was moved to emotion many times reading her failures and her triumphs. I continue to be inspired by the woman in this story who chooses to take action and make changes. Be prepared to finish this book with a newfound motivation to do what is right regardless of what the world may have to say about it.

Meisha Tate
UFC fighter

I've known Mollie for over two decades, and what's always struck me is her unwavering honesty. *Debunked by Nature* is raw, courageous, and refreshingly real. Mollie doesn't just talk about transformation—she lives it. This book is a deeply personal journey that challenges assumptions and invites us to think for ourselves. It's bold, unfiltered, and entirely Mollie.

John Salley
NBA player
Vegan activist

It's often said, "When the student is ready, the teacher will appear." In this case, when the people are ready, the book appears. Humanity is caught in the crossfire of a war against nature. As our world becomes increasingly digitized, the toll on our health and happiness grows ever more apparent. The interconnection between our bodies and the Earth is one of the most intricate, awe-inspiring, brilliant and resilient systems we will ever encounter. Everything we need—food, medicine, healing—grows from the soil beneath our feet. Yet a rising agenda seeks to sever this vital bond with the natural world. *Debunked By Nature* is more than a book—it's a powerful and timely manual for awakening ancient instincts and reactivating the innate wisdom we were born with. It offers the antidote to the disconnection and disorder of our times.

Mikki Willis
Investigative filmmaker

This isn't your typical farm-to-table feel-good story. *Debunked by Nature* is a raw, funny, and gut-honest story of one woman dropping the ideology and picking up a shovel. Mollie Engelhart walks you through her journey from vegan restaurateur to regenerative farmer—and in the process, she calls out the contradictions, faces the death we all try to avoid, and remembers what it means to truly live. This book is a reminder that nature doesn't lie. And it doesn't care about your politics.

Alec Zeck
Founder of *The Way Forward* podcast

As a former vegetarian, I was deeply moved by Mollie Engelhart's *Debunked by Nature*. Her raw journey from vegan chef to regenerative farmer mirrored my own struggle to accept the circle of life. Mollie's unflinching honesty about the realities of farming and her passion for nourishing both land and people through regenerative agriculture inspired me to rethink food, health, and our place in nature's cycle. This book is a powerful call to reconnect with the earth and embrace its wisdom.

Courtney Swan
Realfoodology.com

As someone who nourishes curious young minds each day with some of the most nutrient-dense school lunches in the nation, I witness firsthand how easily children are shaped by powerful food propaganda and the growing pressure to "save the planet" through misguided dietary choices. With profound vulnerability and humility, Mollie shares her personal journey, offering a compassionate, judgment-free invitation to anyone seeking a more truthful path to health. Free from shame or dogma, she gently guides us back to the quiet wisdom of nature, empowering readers to cut through the noise and reclaim clarity. This book is more than a guide—it's a timely and essential conversation. It's a bold call to break free from the myths that have distorted not only our relationship with food, but with Mother Nature herself. I only wish it had found its way into my hands as a young adult—it would have lit the path toward a more nourishing and liberated future.

Hilary Boynton

Founder of School of Lunch

Author of *The Heal Your Gut Cookbook*

"Seek and you shall find" and "The truth will set you free" resonate deeply with Mollie Engelhart's remarkable journey. From celebrated vegan chef to fervent advocate for regenerative agriculture, her transformation reflects a relentless pursuit of truth and a willingness to question long-held convictions. Her story will inspire you to examine your own beliefs around food and farming, and will provide inspiration to the point of action.

John Kempf

Founder, Advancing Eco Agriculture

All I can say is "Wow!" I read the book twice. Mollie did such a good job of conveying her journey—and I purposely call it a journey rather than a story. I found it a very thought-provoking read that really captures what so many are feeling and thinking. Her vivid stories of how life's lessons, grounded in nature, shaped and changed her life are amazing. They are lessons that we can all relate to and learn from.

Gabe Brown

Regenerative rancher

Author of *Dirt to Soil*

In *Debunked by Nature*, Mollie Engelhart shares her gripping journey from vegan restaurateur to regenerative farmer, revealing how real ecological stewardship requires embracing life, death, and the gritty realities of working with—not against—nature. This is a highly entertaining and thought-provoking read.

Kelly Ryerson
American Regeneration
@GlyphosateGirl

Mollie delivers a straight-at-you, no-holds-barred life transformation journey. Her stories bring you closer to God and Mother Nature. Thank you, Mollie, for your transparency and your honesty. This is a must read. The world needs more people like you.

Rick Clark
Farmer

I'm absolutely stunned by *Debunked by Nature*. More than a memoir, Mollie Engelhart has crafted a powerful testament to courage and the transformative power of trusting your inner voice even as the whole world seems to be telling you to silence it. Mollie's journey from celebrated chef to regenerative farmer inspires us to embrace discomfort, challenge our assumptions, and ultimately reconnect with nature's—and God's—undeniable wisdom. She calls us to rediscover nothing less than "the raw, pulsing thread that ties us to life itself." This book is an inspirational tour de force.

Jan Jekielek
Senior Editor, *The Epoch Times*
Host of American Thought Leaders

I've known Mollie Engelhart for over twenty years, and *Debunked by Nature* is the most unapologetically raw and transformative thing she's ever created—and that's saying something. This book is a firestarter for the soul. It doesn't just challenge the orthodoxies of the left and the right—it dismantles them with grace, grit, and the kind of honesty that only comes from lived experience. From her days as a militant vegan to becoming a regenerative farmer and spiritual seeker, Mollie's journey is wild, real, and deeply human, inviting you to rethink everything you thought you knew.

Justin Bua
Artist

Regardless of where our political or ideological differences lie, I believe the reclamation of instinct is essential to our survival—not just as individuals, but as a species. *Debunked by Nature* traces Mollie's transformation from progressive entrepreneur to outspoken activist. What began as a personal unlearning became a deeper reckoning: a confrontation with comfort, a surrender to the soil, and an invitation to question. We may live in a society that lacks curiosity, but this book reminds us that questioning shouldn't be viewed as an act of disloyalty but instead one of survival.

Shihan Van Clief
Poet, filmmaker
Cofounder of Da Poetry Lounge

Mollie's book is a mirror, reflecting truths we've buried under propaganda and noise.

She uses nature—raw, unarguable—to cut through it all. From the vegan movement's one-size-fits-all dogma to equity mandates that ignore common sense, Mollie points to what's real: calves thriving on their mothers' milk, soil reborn under grazing animals, nature's balance over forced sameness. She shows how propaganda can blind us and how nature, with its unyielding truths, sets us straight. She's not asking us to save the world—she's asking us to belong to it.

A.J. Richards
Fifth-generation rancher

Debunked by Nature is a splash of ice water across your face, waking you up to reality and washing away years of lies about the food we eat and how nature really works.

Derrick Josi
Dairy farmer
Author of *An Industry Worth Fighting For*

What if everything you believed about food, farming, and freedom turned out to be upside down?

In *Debunked by Nature*, chef-turned-farmer Mollie Engelhart pulls back the curtain on the echo chambers of modern ideology and shares how Mother Nature herself shattered her most sacred assumptions. Raised on progressive ideals, steeped in veganism, and fueled by righteous activism, Mollie set out to heal the planet. But instead, it was the land—raw, unfiltered, and undeniably real—that healed her.

From LA kitchens to regenerative soil, from plant-based dogma to pastured truth, from abortion advocacy to the sacredness of life, Mollie's journey is nothing short of radical. Not in the political sense—but in the soul-deep, roots-in-the-ground, burn-it-all-down-to-rebuild-it kind of way. It's about choosing stewardship over slogans, humility over hype, and the inconvenient beauty of a worldview that doesn't fit neatly in anyone's camp.

This is not a book about converting sides. It's a story about reconnecting with what's real.

Luke Storey
Author of *A Horse Named Lonesome*,
Host of *The Life Stylist* podcast

debunked

by nature

How a vegan-chef-turned-
regenerative-farmer discovered that
Mother Nature is conservative

MOLLIE ENGELHART

foreword by Joel Salatin

DEBUNKED BY NATURE

PO Box 351
Viroqua, WI 54665 U.S.A.
512-892-4400 • info@acresusa.com • www.acresusa.com

Printed in the United States of America

Front and back cover photos courtesy of Aurelia D'Amore

Publisher's Cataloging-in-Publication

Mollie Engelhart, 1978-
Debunked by Nature / Mollie Engelhart
Viroqua, WI, Acres U.S.A., 2025
156 pp.
ISBN 978-1-60173-598-0

To Miranda "Mimi" Moss
I am forever grateful for your unwavering faith in me.

Table of Contents

Acknowledgements

To my extraordinary husband, who has been willing to live this life with me—I know it hasn't always been easy. There have been moments when you've asked, "Why do we always have to be led? Isn't there an easier way to do things?" Thank you for choosing to walk the harder path with me—the right way—even when it's difficult. Your willingness, your strength, and your partnership mean everything to me. Thank you for our beautiful children and the wild, meaningful, and full life we share.

To my parents, Matthew Engelhart and Jeanne Angelheart—thank you for instilling in me the courage to question everything, to resist following the crowd, and to never blindly accept the way things are. Your guidance shaped me in ways I'm still discovering.

I was not raised by just one or two people—I was raised by a community. To my mother's identical twin, Virginia, and my stepmother, Terces, thank you for your love, your presence, and your part in shaping who I am. To my father's brother, Scott, and to the many others who helped raise me—you know who you are. I carry your influence with me every day. To my brother, Ryland—you've walked this life with me more than anyone else. Thank you for starting me on this beautiful journey of regenerative agriculture. And to the three siblings who blessed my life through my father's second marriage, thank you.

And to my chosen family, who have been willing to work with me, follow me, and support my dreams, thank you. Your loyalty and belief have been the wind beneath so much of what I've built. The truth is, no one does anything alone. You all can take as much credit for all of my success as I can.

My life has never followed a straight line. Like in nature, everything and everyone has a role—a purpose—and nothing thrives in isolation. The roots feed the tree, the tree shades the soil, the mycelium connects and communicates beneath it all, and even the smallest insect plays its part in the balance of the whole. In the same way, every person in my life has held a thread in the web that supports me. None of us can do any of this alone—not truly. We are meant to lean on, lift, and nourish one another. I am who I am because of the ecosystem of people who have surrounded me.

Even in today's world, where everything feels transactional and efficiency often replaces intimacy, we're still part of a community—even if we don't always recognize it. Just because you don't know the Amazon driver who delivers your packages, or the person who stocks your groceries or fixes your plumbing, doesn't mean you don't rely on them. We all need each other, whether or not the relationships are personal. That's the nature of being human. But I have been deeply blessed to experience something more—true community that is chosen, not assigned by transaction. A circle of people who have shown up not because they're paid to, but because they love, support, and believe in one another. That kind of community is rare, and I don't take it for granted.

I thank the Lord for this life—for its beauty, its heartache, its grace, and its lessons. I am only who I am because I reflect the people who have walked with me.

To my employees—many of whose sacrifices I may never fully know—thank you for investing your time and energy into building my dreams. To my neighbors, my loved ones, and most of all, my dear friend Mimi Moss: without you, I would be nothing like the woman I am today. Your belief in me allowed me to believe in myself. Sometimes the greatest gift one person can give another is faith.

Foreword

Many years ago I created a moniker to de-box myself from the liberal environmentalist organic farming stereotype with "Christian-libertarian-environmentalist-capitalist-lunatic farmer." It stuck.

Often shortened to just "Lunatic Farmer," it captured the tensions around being a conservative ecological farmer. I chafed under the assumption that as an "organic farmer" I was opposed to private and home schooling, small government, and the sanctity of life.

To my conservative and faith-based fundamentalist friends, I was a sellout to dominion and capitalism. To my liberal friends, my faith made a mockery of protecting trees, soil, and water. As a child, my church friends made fun of compost and my eco-farming friends made fun of God.

This tension spilled over into political policy. As a teenager during Vietnam, for example, our family saw through the charade and deeply opposed the war. But we did not burn buildings, march in the streets, or fight police. We honored our troops but despised how the elite used them for nefarious agendas.

I was fortunate, I think, to grow up in this dichotomy, but most people don't. Most people don't live with a foot in both camps; most of us choose one camp and vilify the other. If you start a chainsaw for any reason, you hate trees. If you eat a beef burger, you're a murderer. In conservative circles, if you oppose chemical fertilizer, you're a woke snowflake tree hugger earthmuffin commie pinko. In most churches, daring to question styrofoam plates for the potluck dinner lands you in judgmental emotional purgatory.

Rather than wrestling with any inconvenient issues, people tend to silo into their camps, seething and stewing about the hypocrisies and evil of those who disagree. This makes conversion stories incredibly dramatic. This book offers that story in spades.

I question numerous positions I held as a youth, even though I grew up in this wonderful eclectic blend of ecology, faith, and politics. Our family's farm friends throughout the 1970s were primarily hippies. Our faith friends couldn't buy TV dinners fast enough and lobbied for farm subsidies to create mountains of cheap corn. After all, this was a sign of American agricultural might to go along with our military might.

Young people, inexperienced and untried, tend to float along with the agendas cultivated in their families and encouraged by their friends. But maturity brings reflection, and age generally mellows out youthful radicalism. I know it has for me. When we move into a house supported by wood, we realize the chainsaw made it affordable. When we have a child, we realize how special and precious life is. When we encounter extremely capable and sharp people from other religions and ethnicities, we realize prejudice is stupid.

Moving from militant veganism, abortion, and wokeism to beginning to question these positions, and then flipping clear to the opposite side, in one life and one person, makes for quite a revelatory story. Mollie Engelhart slaps a ton of lightweight conventions and assumptions with the unforgiving hand of hard-nosed reality in this fabulous book. I don't know when I've read a book this fast.

I laughed one minute and cried the next as Debunked by Nature vividly narrated the trials and tribulations that lead to a more mature and realistic understanding of life. When collated into this one book, you get the impression Mollie has lived about three lives before turning fifty. But that's the compelling and shocking element of the story.

Cutting her teeth on the militant vegan agenda, going through a couple of abortions, starting her own farming journey, and all the drama in between, Mollie connected to reality, and that changed her faith, politics, and dreams. Bluntly—shockingly so—honest, and written in a sometimes choppy, aggressive style, this book will challenge the most ardent liberal crusader to think twice about what "I know."

Mollie invites us to discover a bigger world, a broader platform, to grow up and understand. And it's all done without losing the desire to honor God's earth, to respect everyone in it, and to build a successful business with hard-scrabble capital. Nearly all the successful people I know got there by working longer, thinking harder, and caring more deeply than the average person. They weren't born with a silver spoon. They earned their reputation, wealth, and success.

Myths and assumptions about business, food, and babies explode off these pages like combining artillery with orchestral harmony. It's beautiful and raw, which makes it compelling, challenging, and wonderful. Everyone who cares about combining earth care with capitalism, combining kids with work, and combining faith with failure will be refreshed and energized after reading Debunked by Nature. This is a ten-course verbal meal. Enjoy.

Joel Salatin
Polyface Farm

CHAPTER I

Return to the Land

L ife was great in 2013.

I was pregnant with Rio, my first, and neck-deep as the owner and head chef of two Sage Vegan Bistro restaurants—one in Culver City, California, and the other in Echo Park. Waiters were humming around, plates were clashing, oat-milk lattes were steaming. Each restaurant was hauling in $5–7 million a year. The two other restaurants I would later start were still years off—Pasadena in 2017,

Me—"Chef Mollie"—in about 2017, at the height of the success of Sage Vegan Bistro.

In 2014 I was pregnant with Rio, getting ready to take our new food truck to Coachella and other music festivals.

Agoura Hills in 2020, ever-so-conveniently in the midst of the pandemic.

But back then, it was just these two—my lifeblood. Customers flooded in—yoga moms, techies, influencers tagging #SaveThePlanet while they ate from their compostable bowls. "Mollie, you're the real deal!" they'd proclaim. "Thank you for this oasis—for feeding us right!" I'd soak it up—damn straight, I'd cracked it. The businesses were raking in cash and changing the game. Hybrid cars were purring outside, paper straws were everywhere, reusable bags were stacked. I believed it all—I was doing the planet's best, one plate at a time.

But a shadowy feeling stuck in me. Their praise fired me up—I'd nod proudly—but I couldn't shake a certain dread. No matter how many boxes I checked—local farmers, organic everything, nutrient-dense food—I'd lie awake, sure we were racing toward a fiery end. The planet was going to sizzle; my wins were just flickers in a blaze.

Customers called the restaurants a refuge. And I'd think, "Maybe . . . but it's not enough." I'd built this business—two spots, big money—yet it felt like shouting into flames.

I wasn't doubting the mission—I was all in—I was just haunted that it wasn't enough to save us.

My back- and front yards in the Knollwood Country Club Estates were my favorite places to be during that time. They were the only places where I found peace. The San Fernando Valley was clipped lawns and HOA rules all around, but my food-forest sprawled. Figs, citrus, avocados—over 100 trees total—a huge vegetable garden, chickens clucking, a compost pile steaming. The HOA fumed. Letters stacked up and fines loomed for my "wild" trees, "noisy" hens, and "rank" heap. I'd smirk, hoe in hand, defying their sterile turf.

This was real—food from dirt, not their golf-course polish. But even there, tending my forest, the dread clung: we're still doomed, no matter what I grow.

Then, one night in 2013, my phone buzzed. It was Ryland, my brother, voice electric. "Mollie, it's soil—soil!" He'd spoken at a conference and watched a panel that included five scientists and an Australian farmer named Graeme Sait. The question cut: "Can humans sustain life here?" The four scientists said no. Collapse was certain. It's all but over.

Me in the crane, putting up the decals for the grand opening of our Culver City restaurant.

But Graeme Sait said yes, flipping cold everything we'd bought about "sustainability." I found his TED talk—Rio asleep in his car seat on the counter, kitchen dark, me in work clothes, smelling of restaurant grease. Sait hit hard: "Soil is alive—the carbon cycles—methane is from food waste, not cows." In the space of just fifteen minutes, my entire world was spinning.

Then it crashed. There were hundreds of pounds of food waste daily from my two Sages—kale stubs, cauliflower butts, avocado pits—mountains I'd overlooked. It could feed soil and rebuild life. This wasn't "save the planet" noise. It was hope—a future I could dig into.

From 2013 to 2018, I hunted for a way to keep my food waste in the loop—turning it into soil, not putrefying into methane in landfills. Back then it was illegal in LA to pay someone to take "waste"—compost—from restaurants. Years later it'd be law to compost—I guess I was ahead of my time. Every time someone hit me with "Mollie, you're my hero!" or "This place is gold!" I'd take it as my shot to pitch them: "Start a farm with me—waste to soil, real change!" From the coach of the Clippers to movie producers to vegan celebs,

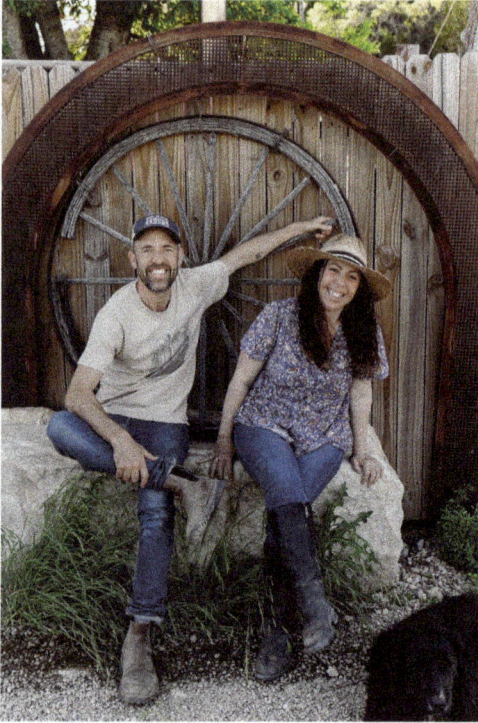

My brother Ryland provided the spark that lit the fire of regenerative agriculture in my life.

they'd nod—"Cool dream"—then fade away. I'd corner my staff too, waving pics of avocado ranches, rambling about microorganisms and Korean natural farming till they'd slip away.

Every night, Rio nursing, I'd scour Zillow—imagining, obsessing, tracing plots, heart racing. I thought the rich could buy a dream like this—I couldn't. My restaurants had big cash flow—$10–14 million total—but profits? Slim: 10 percent, split with investors. I'm not farm-rich, I'd tell myself. And my husband's undocumented status killed deals. Ventura—escrow open, "Yes!"—gone. Moorpark—"Ours!"—nothing. We had three flops in five years of grinding to try to find a farm of our own.

Then, one day, I stopped searching for investors—for the vegan influencers and celebrities who couldn't see it. I'm the one I'm waiting for, I realized. I've got to do this myself.

CHAPTER 2

A New Reality

In 2018, after five years of chasing a farm, we finally did it—we bought twenty acres in Fillmore with Sespe Creek winding through. While my paradigms on a number of issues had already changed, I had no idea what other sacred cows would begin to fall as a result of moving to the farm.

On paper, it was a dream: compost steaming, food waste turned into life, kids running barefoot through trees. But those first weeks were a gut punch—some of the hardest of my life. My best friend, Mimi, was dying of cancer.

At the same time, we had five food setups running at the Coachella and Stagecoach music festivals. I was stretched thin, trying to hold together work, motherhood, and a grief I couldn't even begin to process.

It was the Saturday of Stagecoach weekend when Suze, Mimi's mom, called me crying. "I think she's dying," she said. "I'm calling an ambulance." I was in the RV breastfeeding Luna, dust still clinging to my skin. I radioed my husband: "I need to leave." I packed in 15 minutes—Rio pulling at my leg, Luna crying, Lupe our nanny helping load the bags—and took off in the F-350, sobbing most of the four-hour drive back.

The ER smelled like bleach and panic. Mimi was still in a hallway bay, not yet admitted. Her voice was faint, her hair just barely growing back. "I'm hap-

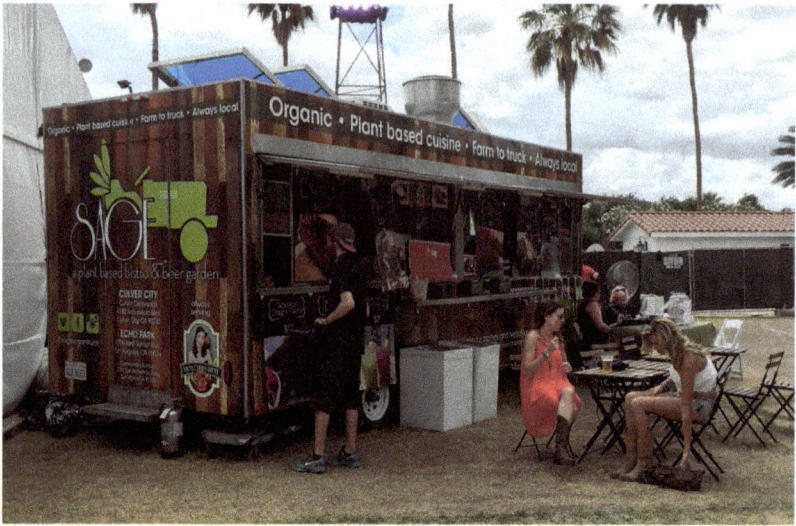

The Sage Vegan Bistro food truck at Coachella in 2014.

py you're here," she said. Those were the last full words she ever spoke to me. I sat with her for hours—kids climbing on me, nurses rushing past, watching her slip away. My chest ached. My best friend, the person who saw me fully, was disappearing.

Mimi and I had built Kindkreme together—a vegan soft-serve shop that eventually became Sage Vegan Bistro. We'd spent 10- to 12-hour days side by side, running businesses, brainstorming, bickering, laughing, crying. She'd once been engaged to my brother Ryland before he left for San Francisco, and in his absence, she became mine. She believed in me more than I believed in myself. And now there was nothing I could do.

Suze and I knew she didn't want to die in the hospital. We pushed hard to get her discharged—argued with doctors, signed paperwork, arranged the ambulance. We brought her home, got a hospice bed by the front window, and called everyone who needed to say goodbye. Friends came in waves, one after the next. We'd been so focused on getting her home that when she finally passed—surrounded by me, Nafeesa, Suze, Bowie, and Bowie's father—we didn't know what to do next. We hadn't gotten that far.

Nafeesa and I held her dead body while Suze gently cleaned her. I was googling "coroner Echo Park" in between tears, thinking, Of course we're doing this now—this is so Mimi. My husband took the kids back to Fillmore. I stayed, helped get Bowie to bed, and held Suze as they wheeled Mimi's body out. We stumbled through shock, trying to hold space for the weight of her absence.

That week was a haze. ER lights. Funeral paperwork. Her house suddenly too quiet. Meanwhile, farm deliveries started arriving—flatbeds with sheds, lumber, tractors. I was grieving and fielding signatures at the same time. The day after she died, Ryland and Sarah had their baby shower. I brought my kids and Bowie, my face tight from crying. Nobody knew

Mimi Moss—my business partner and best friend—being interviewed at the opening of our Pasadena restaurant.

how to act. People skirted around us, small-talking, unsure of how to touch death. I felt like I was floating—there, but not there.

The only thing I could do was throw myself into the land. The house was a mess—cracked walls, sagging floors, dry rot in every corner. We stayed in Granada Hills for another month while my friend Mario helped make it livable. Every dime was gone—poured into the deed. I found sinks on Craigslist for $20 and scavenged pipe from old fields. My husband muttered constantly about the rocks—"too much rock, no soil"—but I couldn't let go of the vision. I saw something in every busted window, every weed-covered slope.

We'd strap the kids in the truck, load it with lumber, and drive the farm together—dust flying, Rio singing, me sketching compost locations in my head. I was juggling three restaurants, prepping a fourth, and trying to breathe life into a place that looked like it had been forgotten.

That first night driving from Sage to Fillmore, Highway 126 felt endless. As Luna slept in her car seat and Rio chattered beside me, I felt panic creep in. What the hell had I done? No backup plan. No money left. Just this plot of dirt. Then I saw them—Trump-Pence signs lining Grand Avenue, Amer-

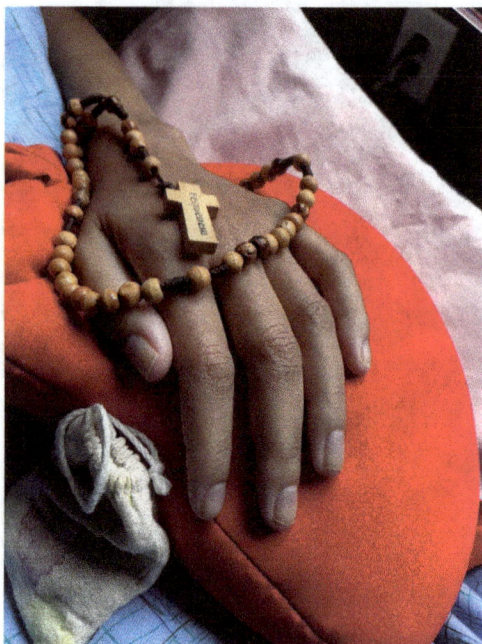

Mimi believed in me more than I did.

ican flags hanging from every porch. I felt like I'd stepped into enemy territory. My husband is Mexican. My kids are brown. Was this even safe?

And then I met Ernie King. His family had owned our ranch since the 1940s and had only sold it in the early 2000s. When he came back from the military, he was given one acre—the one with the original 1884 house on it—smack in the middle of our land. He shuffled around with limited mobility but still showed up every day. "Rip out those trees," he'd bark. "Use urea, forget the organic crap." I ignored that part, but I liked him. He took time to learn my workers' names—Lorenzo, Leo, Jose, Luis, Enrique—and spoke to them with respect. He liked what we were building. He bought a golf cart just so he could come check in with us more easily.

His wife, Patti, welcomed us too. Over time, we became family. I brought them Thanksgiving dinner. During COVID, we did their grocery shopping. That Christmas, Ernie caught COVID—thanks to my husband dropping him at the airport—and gave it to all of us. But we laughed about it, shared soup, and kept moving.

Levi, another neighbor, runs a mulch-spreading company to help farmers reduce herbicide use. We hired him once a year to cover our orchard in wood chips. At my first birthday party on the farm, Levi pulled me aside and asked what was safe to talk about around my "liberal" friends. We stuck to safe topics—home birth, compost, homeschooling—and laughed about it.

That same day, Damion Stehly also showed up. We'd gotten off to a rough start—months earlier, I had made a comment to him about "kids in cages at the border." He laughed in my face and told me there were no cages, just media-fed liberal propaganda. I was furious. But we kept talking. Eventually we loved each other like family. That day, he gave me a ring he made from a quarter stamped with my birth year. One side says "1978," the other "Liberty." I still wear it every day.

I came to Fillmore expecting a fight. What I found were Christian homeschool moms ordering from Azure Standard like I did, families raising their own food, people committed to health and freedom. There were Trump signs, sure—but the people weren't villains. They were kind.

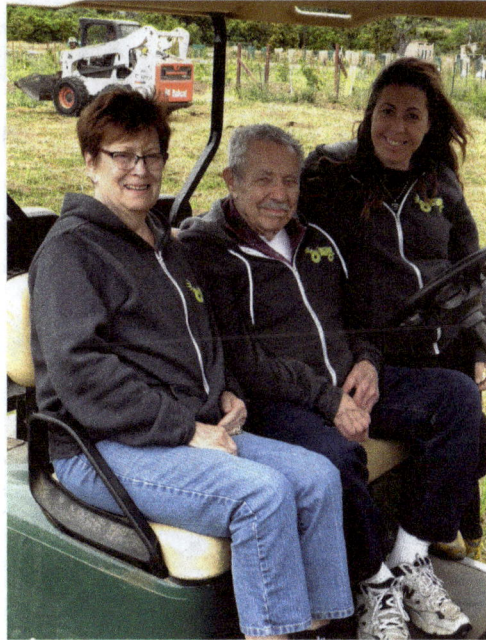

My neighbor and friend, Ernie King, with his wife, Patti. We're all wearing Sage Vegan Bistro sweatshirts—something they wouldn't have been caught dead in before meeting me. Our friendship is such a beautiful representation of what can happen when people simply get to know one another.

They were showing up for their land, their kids, and their neighbors. The story I'd believed about who these people were just didn't hold.

Meanwhile, months into the pandemic, my friends from the city began to turn on me. They questioned whether I cared about people or humanity because I wasn't willing to take an experimental mRNA technology the media insisted on calling a vaccine. But I hadn't changed. I'd never been vaccinated. I was born that way. My kids were too. Suddenly, though, I was dangerous. Selfish. A threat.

That's when I started watching Tucker Carlson—not out of curiosity or rebellion, but because he was the only mainstream voice speaking up for people like me. The unvaccinated. The skeptical. The ones who had questions and didn't want to be coerced. I kept asking myself, Why is he the patron saint of people like us? I didn't want it to be him. I wanted it to be someone else—

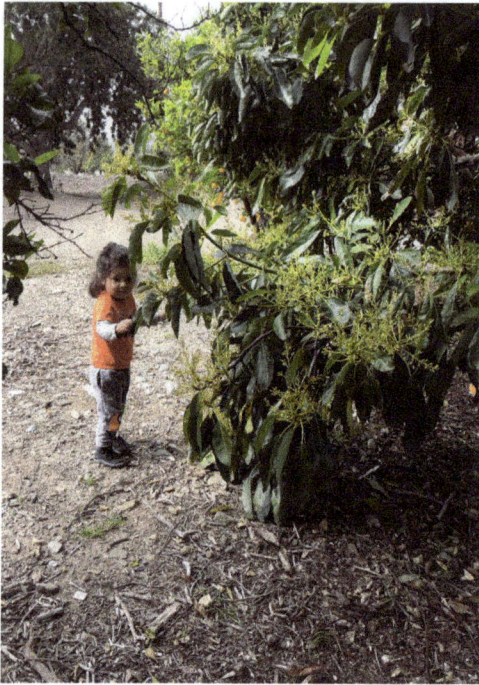

Rio on our new farm by a flowering avocado tree.

someone I already had a favorable opinion of. But it wasn't. It was him. And the more I listened, the more I realized he wasn't the enemy.

He was asking real questions. He was telling the truth about things no one else would touch. He talked about the land, about food systems, about families being broken by mandates and corporate greed. He sounded like someone who actually loved this country, not someone trying to sell it off or burn it down. He was a father, an environmentalist, a concerned citizen. And by the time trains were derailing and chemicals were pouring into huge swaths of America's water supply, he was the only one saying anything.

And I was confused. Weren't the Democrats supposed to be the ones who cared about the environment? About clean water? About corruption and corporate control? But all I heard was silence. Just crickets from the side I thought was mine.

By 2021, I was calling myself a radical centrist—but the truth was, I'd moved even further right of center. Still, I liked the phrase. It made people pause. In a world where everyone was yelling about radical leftists and radical right-wingers, I was standing there with dirt under my nails, trying to grow food and raise free children, not waving a banner for anyone.

The deeper I got into the land, the more the noise fell away. Mimi's death cracked me open. The soil kept me grounded. And the community I once feared became a family I didn't know I needed.

I thought I'd lost everything. But that loss was actually the beginning of the life I was always meant to live.

CHAPTER 3

A Life Less Straight

My path through life hasn't been a straight shot. It's been a winding, unpredictable journey—one filled with lessons carved out of love, faith, and family. I've stumbled, soared, crashed, and rebuilt, each chapter leaving me with a deeper understanding—of myself, of resilience, of what matters.

This is the story of how I got here, from a relaxed baby staring into space to a woman who's weathered rebellion, heartbreak, and reinvention, all while holding tight to the roots that shaped me.

A Chill Baby in a Communal World

I was born to very young parents in upstate New York—my mom was just 20, my dad 21. They were kids themselves, figuring it out as they went. For the first couple of years, we lived with my mom's mother in Connecticut, a cozy, crowded arrangement that gave them a foothold while they found their bearings. Two years later, my brother Ryland came along, completing our little family unit.

The stories from those early days paint me as the chillest baby imaginable. My dad, a house painter by trade, would haul me along to job sites and plunk me at the base of his ladder, and I'd just sit there—napping, waking, staring

I was a pretty happy kid in 1980.

off into space for hours, then putting myself back to sleep. Legend has it I was so laid-back I didn't bother learning to walk until I was past eighteen months old. My parents still marvel at how easy I made it for them.

When I was five, everything shifted. My parents scraped together around $30,000—real money for them back then—and bought an old farmhouse on twenty-seven acres just outside Ithaca, New York. It wasn't just us moving in, though. My mom's identical twin sister, Virginia, and her husband—my dad's brother—came too. The four of them, plus us kids, turned that creaky house into a communal experiment. They shared one wallet and one bank account, pooling every dime they earned. My dad did carpentry, my uncle raised dairy calves, and my mom and Virginia stitched dresses to sell at the farmers market while working shifts at the local co-op and an organic bakery. Money was scarce, but we never went hungry. My mom would get flour from the bakery—sometimes riddled with bugs—freeze it, sift it, and transform it into pies, breads, and desserts that filled our home with warmth. We bought bulk grains and staples with my aunt's co-op discount, and our table was always laden with good, honest food.

That land was our playground and our pantry. We had a massive garden where we grew vegetables—beans, tomatoes, squash—and an apple orchard edged with cherry and peach trees. My childhood was spent barefoot and free, harvesting berries, climbing trees for fruit, swimming in our pond and nearby creeks that crisscrossed the property. We had no television—no distractions from the world outside our little bubble.

My mom and Virginia were vegans, a choice that shaped our meals—except for one quirky twist. One night, my mom left margarine in the oven after a cookie-baking session, and by morning it had melted into a plastic-like sheet. She took one look at it, shrugged, and declared we'd switch to butter. So we grew up vegan, but

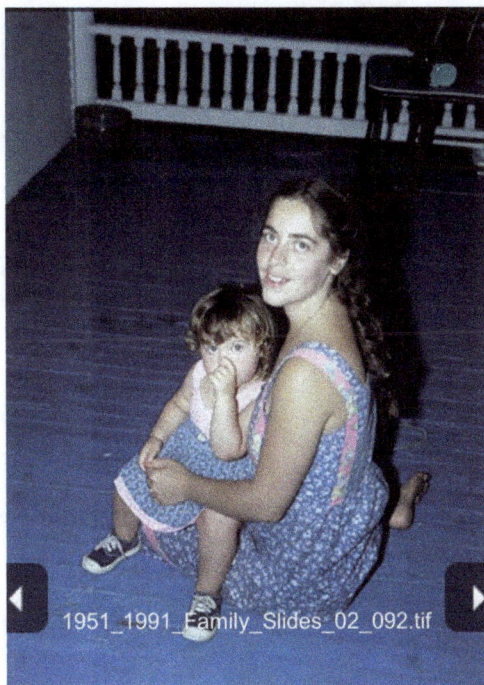

1951_1991_Family_Slides_02_092.tif

Me with Virginia, my mother's twin.

with butter—a small rebellion against dogma that I'm still grateful for.

Freedom, Fumbles, and the Music Hustle

My parents gave me a long leash as a teenager, and I took it and ran—straight into trouble. I was a terror; no sugarcoating it. I sold marijuana, picked the absolute worst boyfriends, and ignored every ounce of good advice that came my way. My folks worked long hours, leaving me and Ryland to fend for ourselves a lot of the time, and I split my energy between responsibility and recklessness. On one hand, I made sure we were fed and got to school every day—highly responsible stuff. On the other, I was a rebel, craving attention, desperate for them to notice me amid their grind.

By the time I hit college, though, the wild streak had lost its shine. I'd already experimented with alcohol and drugs in high school—my parents' lack of strict rules meant I'd been there, done that. It wasn't exciting anymore. I headed west to the California Institute of the Arts, where I majored in photography and minored in film and video. College was a fresh start, a chance to channel my energy into something creative.

On my pony, Smoky, with my brother Ryland and our dog, Jasper. This was in upstate New York in about 1983.

I had a boyfriend there, an aspiring rapper with big entrepreneurial dreams, and I threw myself into helping him. We lived in a condo complex in Santa Clarita, and I struck up a friendship with our neighbor Elizabeth, whose husband, Max Gousse, worked in the music industry. I hustled to get my boyfriend's demo CD into Max's hands, hoping for a deal with Sony Epic. Max wasn't impressed with the album—Chucky's rhymes didn't cut it—but he saw something in me. My grit, my hustle. He offered me a gig as his personal assistant, and just like that, I was in.

Working for Max was a crash course in the music industry. I drove artists to and from the studio and stayed up into the wee hours running errands, soaking in the chaos and creativity. I loved it—the hustle, the late nights, the electric buzz of making something out of nothing. Eventually, I took a leap and opened my own recording studio, first in Hollywood, then in North Hollywood. My dad helped me buy a building—reluctantly, I think, because he and my mom were mid-divorce and guilt might've nudged him along. My brother Ryland and I, who were doing this project together, were clueless, underfunded, and in over our heads. I traded childhood baseball cards for plumbing work, sold my condo to bankroll the building, and, when the money ran dry, moved my bedroom set to the office and slept there, paying the kind Mexican family across the street to use their shower. I'd shuffle over in my robe each morning, bleary-eyed but determined.

Thanks to the connections I'd built with Max, though, business kicked off fast. We had urban legends rolling through—Nas, DMX, Mary J. Blige, B2K, even gospel acts like Mary Mary. The budgets were fat, the vibe was shady, and the kickbacks were plenty. For a couple of years, it was golden. Then the industry imploded. Downloading music gutted sales, and Pro Tools made our fancy SSL board—a beast I was still paying a mortgage on—obsolete. Budgets dried up, and our studio crashed hard. Those years were a blur of trauma,

In the mid-2000s I did poetry slams, including a performance on HBO's Def Poetry Jam.

too—a boyfriend who faked a prison stint to embezzle from me, sexual assaults, business betrayals. I was on the cusp of managing a girl group set to tour with Ludacris and Kelly Rowland when it all fell apart, leaving me gutted but wiser.

When the music industry tanked, Ryland and I had to pivot. We owned a house by that point, and we fought to hold onto it—Ryland hosting at Follow Your Heart restaurant, me touring with spoken-word poetry, even landing on HBO's Def Poetry. We were broke, surviving on leftover soup and bread, renting out every corner of the house to friends. Walk-in closets became bedrooms; the garage housed our buddy Babu; a hallway sheltered half a failed girl group. At one point, we had an entire family—father and three sons—in one room, plus stragglers like my brother's ex and her kid in an outbuilding. Taco Tuesdays and shared TV nights kept us sane. We'd pool grocery money —four bucks here, a can of beans there—and make it work.

Highs and Lows

Eventually, we lost the house. My dad, who'd cosigned the studio lease, in-

I was the self-proclaimed very first gangster vegan.

sisted we sell to settle the debt. He'd flipped the studio building out of foreclosure himself, pocketing $70,000, which stung when he wouldn't budge on the house. "That's a grown-up problem," he'd say when I begged for help—like the time our power got cut off. He was right, though—I always found a way.

The buyer was Stephanie, a successful marijuana grower awaiting prison time. Her husband was already locked up, and they'd worked a deal with the judge to stagger their sentences for their son, Max. Desperate, I asked if she needed a personal assistant. She did—and she bought all my furniture, too. So there I was, the "help" in the house I'd fought to keep. Humble pie doesn't get much thicker. I rented a place nearby, learned the marijuana trade from her, and teamed up with hometown friends Tim and Trevor Ryan—softball buddies who'd stuck by me when the studio folded. One grow house became three, then more. For a decade, I thrived in the gray zone—illegal federally, legal in California—trimming, consulting, building grow rooms. Money flowed fast, and I saved what I could, though it slipped through my fingers just as quickly.

I married Tim, one of my partners in the biz. Seven days after our wedding, the feds raided our Granada Hills house—bought with trimming cash—seizing hundreds of pounds of weed and stacks of money, only to drop the charges for "insufficient evidence." Jack-and-release, we called it—they take everything and let you go. They hit Tim and Trevor's place too, so we bolted north to a tiny, pet-stinking cabin until things cooled. Trevor grew distant, and then came the call that flipped my world: his attorney called and said Trevor was caught on a plane in Madison, Wisconsin, with 150 pounds. Wisconsin wasn't California—150 pounds there meant prison, not a slap on the wrist. He got five years, and I spent them juggling his mess and mine—saving two of his properties, losing two properties and enduring lots of hardship, including robbery and squatters, hauling weed out of culverts to dodge raids. It was exhausting, exhilarating, and humbling.

Reinvention: Ice Cream, Sage, and Stability

With the kitchen crew at our first restaurant in Echo Park in 2011—including Elias (to my right), who was not my husband at the time.

Amid the chaos, Ryland's ex, Mimi, got pregnant and moved in with Tim and me after her baby, Bowie, arrived. It was a sweet, grounding time—planting fruit trees, tending gardens, just being together. But Mimi itched to provide for Bowie, so we started a raw vegan ice cream shop on Chuhunga Boulevard in Studio City. I'd been making it for years—my dad and Woody Harrelson were fans—and it took off. We opened a second spot in Pasadena, but the niche was too narrow. Then two men, Robert and Tony, walked into our shop with a proposal: partner with their vegan restaurant, Sage, on Sunset Boulevard. We'd handle desserts and drinks, they'd do the food. It sounded good, so we dove in.

Sage opened in March 2011, and Mimi and I were proficient, so Robert and Tony shifted more and more responsibilities to us. They bent vegan values in ways that grated on me, so I bought them out—$75,000 to join, $150,000 to take over. I opened a second location in Culver City, and Sage became my life's work.

Along the way, I met Elias, a dishwasher Robert once praised as the ideal employee. He's now my husband—the greatest gift from that rocky partnership. The past decade has been about growing Sage, building a life, and reflecting on the wild road that brought me here. This book starts where that journey pauses—a moment to look back before stepping forward again.

CHAPTER 4

A Farm Where Nothing Dies

When I became a farmer, I had a big idea: create a farm where nothing had to die. Animals would live out their days munching grass, with no blood on my hands. It was straight out of my vegan chef playbook.

I'd built Kindkreme and Sage on that ethical stance: compassionate food, no suffering, all love. Picture me in LA, dishing out cashew ice cream and kale bowls, proud as can be, thinking I'd cracked the code—I was feeding people while saving the planet, and no harm done. Fillmore was going to be the next step: twenty acres of peace—a sanctuary where every goat, duck, and sheep got a happy ending. I was dead set on it, like it was a mission from God.

It kicked off fast—animals showed up before I could catch my breath. Unwanted goats, sheep, pigs—people's failed pets or 4-H projects were all dumped on us. I already had a parrot rescue going, so parrots rolled in too— squawking, bright, a mess of feathers. I'd stand in the orchard, Luna on my hip, watching them settle, thinking, "This is it—a vegan paradise." I'd spent years perfecting recipes—no dairy, no meat, all plants—and now I'd perfect this: a farm where life just hummed along without slaughter or compromise.

My husband, Elias, wasn't on board, though. He grew up in a village in

Oaxaca, Mexico, where they grew or foraged or hunted all their food. My no-death ideas made no sense to him. They clashed with a life where survival meant balance, not mercy. I waved him off. I'd make it work. I'd pulled off tougher things than this.

Early on, Elias and I planted 1,000 avocado trees—every variety I could get my hands on. I had this dream—an Avocado of the Month Club, shipping out funky ones like Gwen or Pinkerton with recipes from me and

My daughter Luna with one of the Khaki Campbell ducks we raised, before they were all killed by the drakes—which was really due to my arrogance.

him. We'd spent weeks digging holes, hauling young grafted trees, sweating under the Fillmore sun, picturing lush groves. But the land was out of whack—decades of neglect and herbicides had put it totally out of balance. Without this balance, one species will explode.

In our case this meant thousands of ground squirrels, which came out of nowhere, chewing through the roots, killing our trees faster than we could plant. We started trapping—little cages baited with discarded leftover parrot food snapped shut all day long. I talked to a neighbor with a falcon business; he was thrilled to take them. I felt good—this was a natural predator fulfilling the circle of life—it wasn't just waste. Then he said he could only handle three a day. Three. We were catching thousands—sometimes upwards of 50 to 100 in a day—drowned and composted. I knew they were dying; I just pretended I didn't. I'd skirt the trap lines, eyes down, telling myself it wasn't happening. The truth is, to get young trees established, thousands of rodents die. It's a fact—one I couldn't dodge forever.

Then there were the toads—thousands of them, hopping everywhere.

We planted hundreds of avocado and citrus trees and tried—in vain—to protect them from squirrels.

Again, without balance . . . one species will explode. Any rain or moisture turned the place into a toad frenzy. They'd bounce off the porch roof, splatter on the driveway, and cover the ground like a moving carpet. You couldn't drive without squashing hundreds—tires crunching, me wincing, kids shrieking in the backseat. I'd swerve, slow down, and try anything to avoid it, but there was no way around. The imbalance screamed at me—this wasn't the Eden I'd planned; it was nature doing its thing, loud and messy.

One night, driving back from Sage, my vet called—an emergency. A hoarder had died in a San Fernando Valley trailer park, leaving dozens of animals in filth. His sister wanted them saved; his husband wanted them gone. Animal control was circling. I punched the address into GPS and floored it. When I got there, it was chaos—pigs, turkeys, goats, ducks, chickens, and parrots were crammed into a double-wide with no yard—just stink and squalor. Volunteers yelled, family fought, madness reigned. Before I could step out, they started shoving parrot cages into my truck. My husband pulled up with the van, and we packed it with birds, ducks, and chickens—anything we could grab. Back in Fillmore, we worked past midnight, setting up cages and doctoring birds. The ducks started waddling into the orchard. I'd read that ducks ate invasive snails, so I thought, "Perfect—soil helpers and no killing." This was going to work.

The ducks took off—at first. They multiplied fast, sitting on eggs and hatching babies—more than I could count. But they weren't as committed to the snails as I'd hoped. It was less pest control and more lounging. I dug into it and learned that runner ducks or Khaki Campbells were the pros, not these ones. So I ordered babies online—fluffy, peeping miracles shipped to my door. The kids helped—Rio splashing water, Luna giggling, helping change bedding and feed them.

Ten weeks later, they were ready. It was a cool morning as they disappeared into the tall grass, beads of water rolling off their backs and tall blades brushing their sides. I puffed up my chest, thinking, "We're regenerating—vegan farm nailed."

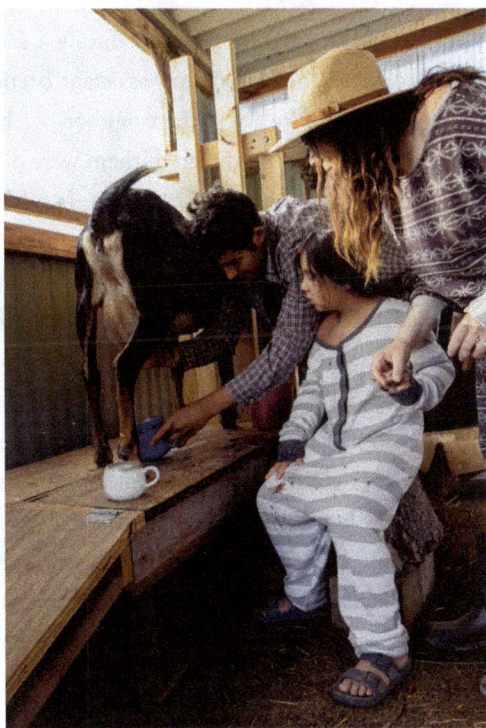

Osmar, Rio, and me milking.

Hours later, my husband called. "Did you check your phone?" I hadn't—my hands were covered in flour in the kitchen. "You're not gonna like it." I rinsed off and opened the messages. All the ducks were dead. Because of my unwillingness to cull the male ducks—because of my vegan mindset—these cute ducks I'd just cared for for the last ten weeks had been raped to death by the older male ducks. It was an imbalance I'd caused, thinking nothing should ever die. Nature didn't care.

I called my father, sobbing. "That's enough to bring a vegan to her knees," he said. That night, I told my husband, "I don't wanna know when or how—just get rid of 90 percent of the drakes." Alyse, my girlfriend who lived with her family on the farm, along with her husband, Dan Doty—a hunter—and my uncle Scott processed them and fed the community. I pretended it didn't happen and clung to my no-death dream.

Then one morning I sat on the porch swing, enjoying the sunrise peeking over the mountains, feeding Luna, when I saw my uncle tear by in the electric cart, face tight. He stopped, yelled for my husband, then grabbed a gun. I corralled the kids and my assistant Brenda and followed, heart pounding—what now? We got to the neighbor's orchard, where my sheep had been grazing, and I froze. Eighteen of them were down, throats ripped out, bodies untouched. Two German shepherds—family pets, not strays—sat panting, tongues lolling, pleased with their sport.

I stood there, shaking. Nature's brutality hit like a slap. Three of the sheep twitched, barely alive; my husband shot them to end it. He sorted fast—some for compost, some for meat—yelling, "Call people—we don't have the freezer space!" He worked 12 hours straight, skinning, cutting, and cleaning, while I dialed. Most Americans said no—"Trauma meat's not my thing." I didn't get it—all meat's got trauma, right? But Latino farm workers and restaurant families came—buckets of lamb were hauled off, feeding dozens. I'd have composted it all and cried; my husband, raised hunting, saw use where I saw loss.

The land wasn't just imbalanced with critters—its soil was denigrated, wrecked by herbicides and neglect. I brought in organic fertilizer, thinking I'd fix it clean. Then I read the bags—blood meal, bone meal, feather meal, chicken and cow manure. I asked the company, "Where's this from?" Consolidated feedlots, they said—factory farms, not organic pastures. My head spun—organic fertilizer was full of pesticide-fed, herbicide-laced animal manure. That's when we got Una, our cow. My father gave her to us, already bred, her first calf a bonus. The calf favored two teats, so we milked the other two.

Another realization came one day while breastfeeding Luna. I felt good about nursing my kids past two, as natural as can be. Rio toddled down. "Milky?" he asked. I grabbed a Costco oat milk tetra pack from the fridge—and it hit me. Una's milk was right there, from an animal sharing our land, our water, our germs. Why was I pouring processed oats into Rio when Una's milk sat feet away? I'd been told dairy was bad, but how could my milk be perfect—divine, natural—and Una's be wrong? That didn't make sense. I started to wonder—what if veganism wasn't healthier? What if there wasn't a way to avoid death—that death was in everything?

Death was happening, always—every moment of every day—and so was life. There's no life without death. There's no way around it. To be alive, to eat, means consuming death.

I saw it everywhere—ground squirrels in traps, toads under tires, ducks and sheep gone wrong. You can't separate from the life-death circle; it's baked in. Every plate has death on it—even in the vegan plate, there might be a squirrel, a rabbit, or something you don't see. I'd been kidding myself, thinking I could sidestep it. No matter what I put on my plate, I'm causing death.

I realized I had to have reverence for this—like there's a dead animal staring back. Because there is—maybe it's a ground squirrel, a rabbit, or the bones and blood of cows from the store. There's no stepping outside that circle.

CHAPTER 5

Masculinity: Holding the Line against Nature's Chaos

I n the city, there's this idea floating around that men don't matter much. You hear it all the time—maybe we need a few for making babies, but beyond that, women can do it all, and better.

Living in LA, disconnected from the land, that sounded plausible. I bought into it for a while. At 36, I was unmarried, running Kindkreme and Sage, killing it in the restaurant game. I'd been married once, sure, but it didn't stick, and I figured I might never have kids. Didn't faze me—I was financially independent, calling my own shots. Men? Nice to have around, maybe, but not essential. I'd built my life without them, and I was proud of that.

One afternoon I was on the patio at Sage in Culver City, venting about men to Pablo, one of my employees. He was a quiet, solid guy. He listened while I went off about how I didn't need anyone. Mid-rant, he cut me off: "You've got nothing to offer a man." I stopped cold. What? Me—successful, driven, important—nothing to offer? I was stunned. He didn't back down; he just laid it out: from his view, rooted in a traditional take on marriage, my worth wasn't in my bank account or my hustle. It was in being there—cooking, raising kids, building a home. I worked twelve-hour days, seven days a week, married to my restaurants. What could I give a man who wanted a partner, not a boss? I brushed it off, but it stuck—like a splinter I couldn't pull out.

Maybe there was something to it.

Less than a year later, I married my husband through extenuating circumstances we'll get into later, and that splinter turned into a full-on sore. Elias is from a village in Oaxaca, raised on the land, where life was growing, hunting, surviving—practical, grounded stuff. He wanted love in ways I didn't get—home-cooked meals, clean laundry—small acts I shrugged off. I'd snap, "Why do you need to be served?" I didn't understand why ordering takeout or bringing home food from the restaurant wasn't good enough. Sure, I'm an excellent chef, but that's what we do all day—cook for people. Why should I come home and have to cook for him?

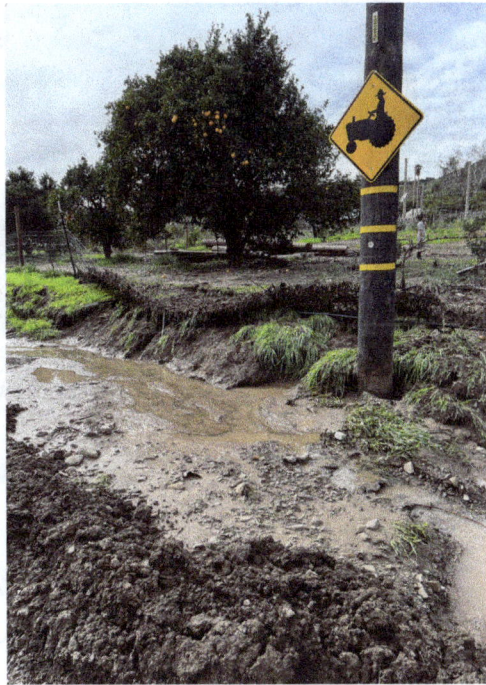

In January 2020 we got eighteen inches of rain at Sow a Heart Farm.

I didn't see it—I didn't want to. To me, love was big moves, ambition—not frying eggs. But he'd look at me, quiet, waiting for something I wasn't giving. I'd get frustrated, thinking he was asking too much, not seeing how deep those needs ran—how they weren't about logic, but about something older, something tied to nature.

Then we moved to the farm, and everything flipped. That city idea—men as optional—fell apart the second we stepped into real work. Fillmore wasn't a place for theories; it was dirt, sweat, and survival. My husband's hands started shaping the land—planting orchards, installing irrigation, turning rocky soil into something alive. I'd design the layout of where everything would go—from animals to compost piles to gardens and greenhouses—and he'd dig the trenches, haul the pipes, and make it happen. I could plan, nurture, and hustle, but when it came to raw strength—to holding the line against nature's

Water and debris rushed across the farm, pulling our propane tank and everything else with it.

chaos—I saw what he brought. It wasn't optional—it was the backbone.

Take the basics. Who builds houses? Who fixes the plumbing when it bursts? Who keeps the power on, lays the roads, hauls the heavy loads? In LA, I'd never thought about it—water flowed, lights flicked on, food landed in stores. But on the farm, I saw the work behind it. My husband would labor in the hot sun when it's over 110, keeping the animals alive—carrying water buckets, fixing fences—while I stayed inside, cool and air-conditioned with the children. He'd dig irrigation lines, muscles straining, while I'd figure out what to plant. I don't know many women who'd take on those jobs—or want to. Even the fiercest among us lean on that labor, even if it's outsourced to some guy we never meet. The idea that men aren't needed only holds up if you're blind to what keeps the world running.

Then came the flood—January 2020—eighteen inches of water roared across our land like a beast we couldn't tame. I was eight months pregnant with Estrella, belly huge, moving slow. It hit fast—rain pounding, Sespe Creek swelling, everything we'd built at risk. My husband, my brother Ryland, my brother-in-law Trevor, and our friend Nicholas Osmar jumped in, with no hesitation. They waded waist-deep into the muck, boots sinking, hauling animals to higher ground, tying down equipment, wrestling a thousand-pound propane tank to keep it from floating off. I grabbed my employee Nicole and Christina, one of our workers, and we did what we could—moving baby goats, pointing out hazards, shouting over the roar. But soon, I had to step back—too pregnant, too slow. I retreated to the house, started a fire, and made bread and beef stew, knowing they'd come in soaked and starving.

At one point, my husband was wrestling the tractor, positioning it to block that propane tank from drifting down the road. I yelled, "Careful—you'll hit my mulberry tree!" He snapped back, "¡Cállate, cabrona!"—"Shut up, boss!" in rough English. A new worker muttered, "Toxic masculinity much?" I laughed to myself. Toxic? No. This was the masculinity I needed—testosterone, strength, the courage to save our farm. I was right where I belonged—inside, warm, feeding the kids, keeping the home steady while they battled the water. That night, employees got stranded—the roads were flooded with no way out. We turned it into a slumber party, kids piled on blankets, adults swapping stories by the fire.

Scrolling Instagram later, I saw a woman in Lake Arrowhead, snowed-in from the same storm, filming herself: "I've spent my life thinking I don't need a man. But if anyone digs me out, I'll marry you on the spot!" I nodded. Nature doesn't mess around. When it hits, you don't debate. You need strength,

Elias and the other men jumped into action and saved our farm from even more destruction. (In this photo, Elias is merely being a wonderful father—we were too busy that day to take photos of his heroics!)

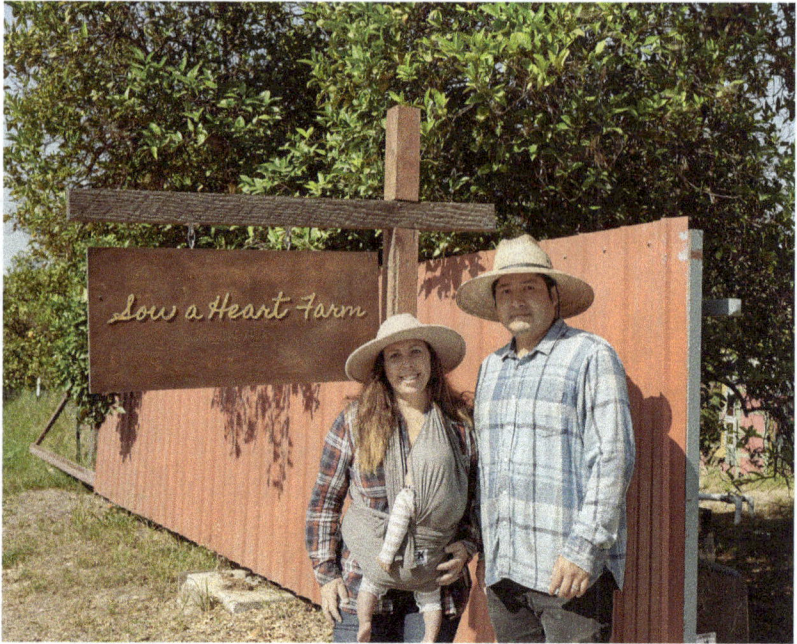

While a decade ago I would have denied it, I know now how much I need my husband's strength—and how much our society needs men like him.

action, someone who'll step up. My husband did, and does—every day, not just in floods. He'd butcher a sheep when dogs tore through it and process it for the community, while I'd just freeze up. He'd fix things when they break and haul hay when the animals needed it. I'd plan the feed, but he moved it. I'd birth kids at home, nurture them, keep the rhythm; he'd plant the orchards that fed us.

Back in LA, I'd have bristled at this—at 36, I'd have said, "I don't need that." But at 46, on the farm, I'm grateful. Grateful for my marriage, for my four kids, for serving my husband and family in ways I once didn't understand. Suffering doesn't come from embracing womanhood—it comes from fighting it. We've got to stop outsourcing masculinity, pretending it's optional just because we can order groceries online. It's not toxic—it's essential—a balance to femininity.

If you think men don't matter, ask yourself: in a disaster, would you want only women and kids around? If no, you already know the truth. We need men—strong men—not just one, but a whole society of them. Without that, we can't stand.

CHAPTER 6

Death, and Our Relationship to It

When I first stepped onto the land as a farmer, I carried with me the heart of a vegan—a heart that recoiled at the thought of death, that saw it as a failure, an aberration, something to be resisted with every fiber of my being.

I had spent years in Los Angeles, building a life around Sage Vegan Bistro, where every dish was a testament to my belief that we could live without harm—that we could sidestep the messy, bloody reality of nature. I imagined a world where nothing had to die unless it was old and ready, slipping away peacefully in its sleep, surrounded by quiet reverence. That was my dream—a sanitized, gentle existence where life could flourish without the shadow of death.

It's a beautiful dream, isn't it? One we all might wish for ourselves, for our loved ones, for the animals we cherish. But nature doesn't bend to our dreams. Nature is raw, unapologetic, and relentless. It taught me, in ways I couldn't ignore, that death isn't separate from life—it's the very thing that makes life possible. The gazelle racing across the savanna doesn't drift off into a painless slumber; it meets the lion's jaws, its blood spilling to nourish the predator and her cubs. In every breath, every bite, something is living because something else has died—or is dying—to sustain it.

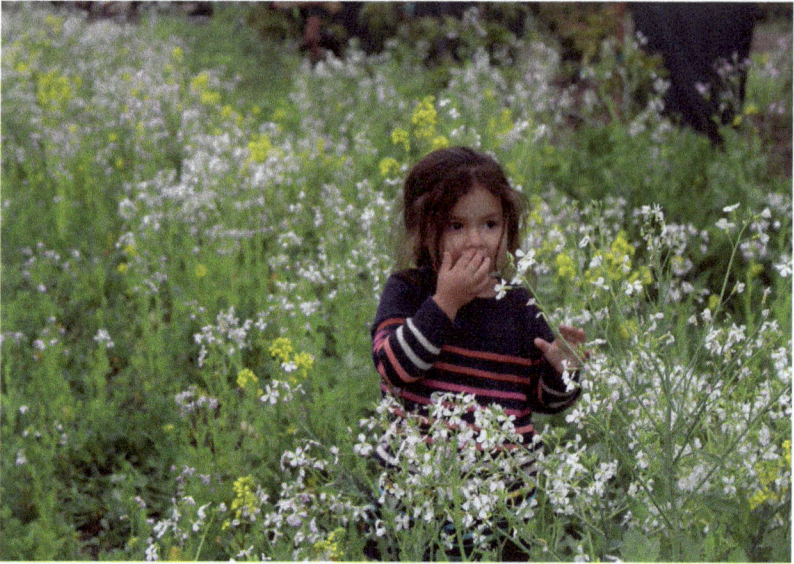

As a vegan, I thought everything in nature was as innocent as a child. Yes, nature is beautiful, but the reality is much more stark.

As a vegan, I thought I could opt out of that cycle—that my choice to abstain from meat was a moral high ground, a way to rise above the brutality. I was wrong. Nature doesn't offer an exit ramp.

Where Do We Draw the Line on Death?

The reality hit me hard when my husband and I planted thousands of grafted trees on our farm in Fillmore. We had this vision—rows of lush, green avocado trees heavy with fruit, a testament to regenerative agriculture, feeding our community with nutrient-dense food.

But, as I described above, almost as soon as we broke ground to plant the trees, the ground squirrels and voles descended. Their population explosion was a ticking bomb of imbalance. I was stuck—caught between my ideals of a death-defying farm and the reality staring me in the face.

We hadn't planned for the fact that those rodents were fulfilling a natural function. With so many of them scampering around, the coyotes lurking beyond our fences paid little attention to my chickens. The poultry strutted freely, safe in the shadow of the squirrel glut. But as we trapped more and more squirrels, thinning their numbers, the balance shifted. Many mornings we would find feathers scattered, hens gone without a trace, taken by the lean and cunning coyotes who had turned their yellow eyes toward my flock.

Nature doesn't negotiate—it adapts. Fewer squirrels meant more pressure on my chickens. It was a lesson in ecology I hadn't signed up for, but one I couldn't escape.

Years later, after relentless trapping, we found a tenuous stability. The grafted trees grew tall enough to withstand some nibbling, their roots digging deep into the soil. We brought more dogs onto the farm—scrappy, loyal mutts who patrolled the edges, keeping both

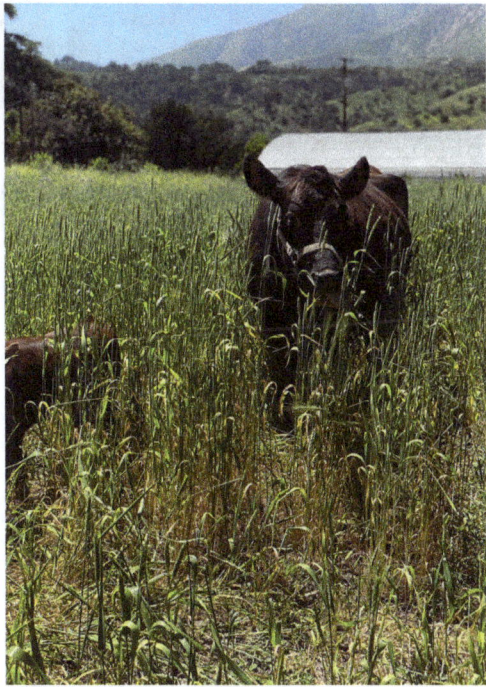

Una, my first cow. She helped me realize that dairy isn't evil—it's a superfood! If human breastfeeding is good, how can dairy be bad?

squirrels and coyotes at bay. And as I watched the land settle, I began to see the squirrels differently. They weren't just a problem to be eradicated. Their tunnels aerated the soil, letting water seep in where it might've run off. They were part of the system—vital, even—until they tipped the scales too far.

Death wasn't the enemy; imbalance was. I wrestled with that truth for a long time. I'd feel a pang of guilt when we drowned the excess squirrels and tossed them into the compost pile, their bodies breaking down to feed the microbes that enriched our soil. But when the falconer took them, when their deaths fueled his birds, I felt peace. Why? Why did one death weigh heavier on my conscience than another? It gnawed at me, this inconsistency, until I realized it wasn't about the act of dying—it was about my need to assign it meaning, to fit it into a story I could live with.

Where did I get the idea that I could create a world without death? Every species fights to survive—it's written into our DNA, into the fabric of life itself. A deer bolts from a wolf, a fish thrashes against a heron's beak, a squir-

Giving chickens access to most of the farm—as opposed to confining them to their own area—brought ecological benefits to all of our enterprises. But it also meant a few casualties to predators.

rel scrambles from a trap. Survival is instinct. But I'd convinced myself, in my vegan days, that I could step outside that struggle, that my choices could re-write the rules. I was na-ive. Nature doesn't bend to human declarations. If a population—squirrels, deer, anything—grows beyond what the land can sustain, something will give. Predators will rise, disease will spread, or starvation will set in. Death is the mechanism of balance—the hand that keeps the scales from tipping too far.

Some argue this is happening to us now—that humanity's infertility and our rising rates of chronic illness are nature's way of correcting our own over-reach. I've heard it said in hushed tones, in podcasts and in late-night con-versations: maybe the Earth is pruning us back. But I don't buy it—not fully. Nature balances, yes, but what we're seeing isn't nature's doing. It's ours. We're poisoning ourselves—dumping endocrine disruptors into our water, spraying glyphosate on our crops, filling our bodies with processed junk that starves our microbiomes. Sperm counts are plummeting, women are struggling to conceive, and we act like it's inevitable. It's not.

This isn't nature's correction; it's our collapse, born of arrogance and a refusal to face the reality we've created. We feel guilty for existing, so we let ourselves be diminished, as if our presence is the problem. Nature doesn't hate us—it's indifferent. We're the ones sabotaging ourselves.

In the wild, a lion doesn't pause to mourn the gazelle it tears apart. A coy-ote doesn't apologize to the rabbit twitching in its jaws. They eat because they must, because survival demands it. There's no guilt, no moral debate—just the rhythm of life and death playing out as it has for millennia. But we humans? We've built an entire moral architecture around death, a shaky scaffold of rules and exceptions that we cling to like a lifeline.

I used to think eating a cow was wrong—cruel, unnecessary—but I'd pluck a carrot from the ground without a second thought. Why do we decide that eating a cow is wrong but eating a carrot is fine? Modern science tells us plants might be sentient in their own way—communicating through chemical signals, responding to threats, even "screaming" in frequencies we can't hear when they're cut. So why do we feel horror at the thought of eating a rabbit but casually eat shrimp—which, by all measures, are highly intelligent and communicative? We've watched shrimp scuttle across the ocean floor on documentaries, their tiny bodies pulsing with life, yet most of us eat them by the handful, dipped in cocktail sauce, without a flicker of remorse.

Babydoll sheep—prior to the massacre.

In nature, these distinctions aren't moral—they're practical. A predator eats what it can catch, what its body needs. Survival draws the line. But we've lost that clarity. We're so far removed from the fight for survival that we've turned death into a philosophical playground. We debate the ethics of meat while sipping oat-milk lattes, ignoring the field mice shredded by combine harvesters to grow our oats. We agonize over factory farms but shrug at the pesticides that wipe out entire ecosystems to protect our kale. Our comfort—our grocery store, our delivery app—has given us the luxury to judge death from a distance, to pick and choose what feels "right" without ever facing the consequences.

And yet, the consequences are there, whether we see them or not. Every bite we take is tied to death—somewhere, somehow. The vegan avocado toast I once served with pride? It came at the cost of those squirrels—tens of thousands of them over the years, their lives snuffed out to protect the grafted trees. The organic spinach in my salad? It grew in fields where insects were crushed, where soil was tilled over nests and burrows. We can't escape it. We've just learned to hide it—to pretend our hands are clean because the blood isn't on our plates.

Death Is Not a Failure—It Is a Function of Life

I used to think death was a mistake, a flaw in the system that we could fix if we tried hard enough. As a vegan, I believed that by refusing meat, I was voting for a world with less suffering, less loss. But farming stripped that illusion away. The sheer scale of death tied to my grafted trees alone—tens of thousands of squirrels over the years—was a reckoning. And that was just one crop, one piece of my diet. Multiply that across every vegetable, every grain, every fruit I'd ever eaten, and the numbers became incomprehensible.

Vegans often point out that most crop deaths come from animal feed—corn and soy grown to fatten livestock. It's true; industrial agriculture is a death machine on a massive scale. But the row crops that make up a vegetarian diet—wheat, rice, beans—aren't innocent. They're sown and harvested with machines that grind through habitats, killing rabbits, snakes, birds, and countless insects. The pesticides we spray to protect them poison waterways, wiping out fish and frogs. You can't wash that off your lentils. You can't separate yourself from the death in your food just because it's plant-based. It's there, woven into every bite.

Everything alive exists because something else has died. It's not a tragedy—it's a fact. A cow grazes on grass, and that grass grows from soil teeming with microbes that feast on decayed matter—dead plants, dead animals, dead everything. The soil itself is a graveyard, a living testament to death's role in creation. Those microbes break down carbohydrates from photosynthesis, turning yesterday's life into today's nourishment. When the cow dies, it returns to the earth, feeding the cycle again. There's no waste, no failure—just transformation.

We'll all die someday, and when we do, we'll feed something else. Maybe worms, maybe bacteria, maybe the roots of a tree that shade someone's porch. That's not morbid—it's beautiful. It's the promise that nothing ends, not really. Death isn't a full stop; it's a comma in an endless sentence. But we fight it, don't we? We cling to life, to youth, to control, as if we can outrun the inevitable. We've built a culture that sees death as the ultimate defeat, something to delay with pills and surgeries and sterile hospital rooms. Nature doesn't see it that way. Nature knows death is the engine of life.

Facing Death, Finding Life

Stand in a forest, a field, a farm—anywhere life thrives—and you'll see it: everything is eating, and something is dying. A hawk snatches a

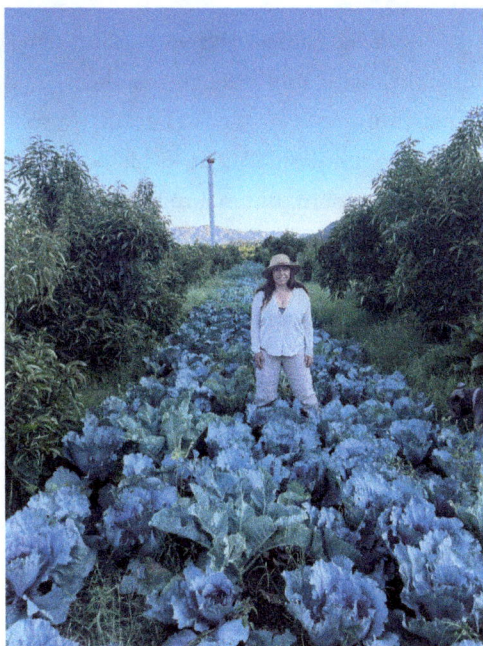

I'm a vegetarian now, but I feed my family and my community meat grown on our farm, gratefully acknowledging that the death of one organism gives life to another.

mouse, a beetle devours a fallen leaf, a root sucks nutrients from decomposing matter. It's happening right now, all around us, in every moment. That's the heartbeat of the planet—life and death pulsing together, inseparable.

I didn't get that as a vegan. I thought I could stand apart, that my choices could carve out a little island of purity in a brutal world. But nature doesn't care about my intentions. It doesn't care about my guilt or my ideals. It just keeps going—eating, dying, regenerating. The sooner I accepted that, the sooner I could let go of the fantasy that death was unnatural. It's not. It's the thread that ties everything together.

Living on the farm has taught me to honor that thread, not to fear it. When we lose a sheep to a dog, we don't let it rot—we process it, feed our community, return what's left to the soil. When a tree falls, we chip it into mulch, letting it protect the ground it once grew from. Every death has a purpose, a place. That's not callous—it's reverence. It's recognizing that we're part of something bigger, something that doesn't end with us.

One of the hardest days on the farm came early on, when those two German Shepherds tore through our sheep. That day cracked something open in me. Nature doesn't flinch at death; it uses it. My husband, raised hunting in Mexico, understood that in his bones. I'd been taught to turn away, to sanitize, to mourn. But mourning doesn't change the reality—it just keeps us stuck. Over time, I learned to look at death straight on. When roosters had to be culled, when a calf didn't make it, when a tree succumbed to drought—I stopped seeing it as failure. I started seeing it as part of the dance.

This shift didn't come easy. It took years of wrestling, of crying into the dirt, of questioning everything I'd built my life around. But it brought me peace. Not the shallow peace of denial, but the deep, steady kind that comes from alignment. We're not here to avoid death—we're here to live with it, to work with it, to let it teach us. The farm became my classroom, and death was the sternest, truest teacher I'd ever had.

Death in a Disconnected World

Our society doesn't get this. We've pushed death to the edges—hidden it in slaughterhouses, hospitals, crematoriums. We eat meat from plastic trays, not knowing the animal it came from. We've sterilized death, made it a stranger, and in doing so, we've lost our grip on life. We obsess over longevity—smooth skin, strong hearts, endless youth—while ignoring the poisons we pump into our bodies, the chemicals we spray on our food, the systems we've built that are killing us slowly.

Nature doesn't hide death. It flaunts it. A carcass on the plains feeds vultures, ants, and soil. A fallen tree becomes a nursery for fungi and beetles. There's no shame, no secrecy—just purpose. We could learn from that. Instead, we've convinced ourselves we're above it, that technology or morality can save us from the cycle. But we're not above it. We're in it, whether we admit it or not.

I think about my vegan days sometimes—how I'd sit in my restaurant, sipping a latte, preaching compassion while blind to the death in my supply chain. I wasn't wrong to want less suffering; I was wrong to think I could escape the cost. Now, I feed my children, my husband, and my community meat from the farm—lamb, duck, beef—knowing exactly where it came from, how it lived, how it died. I don't eat it myself—I'm a vegetarian now—but I don't flinch anymore. I give thanks. I see the life it gives them. I don't think that's

Through life and through death, we belong to something much greater than ourselves.

hypocrisy—it's honesty.

Even the way we as humans die reflects how we think we're separate from nature. In the developed world, particularly in the United States, we bury bodies in boxes that can't break down—steel caskets or polished hardwood, sealed tight against the earth. Those bodies are filled with embalming fluid, a cocktail of formaldehyde and other chemicals, preserving them in a sterile, lifeless state. They're wrapped in synthetic fabrics—polyester suits, satin linings—cut off from the soil, the microbes, the cycle they're meant to rejoin. It's as if we believe we can extract ourselves from the whole, even in death.

The death rituals in the United States speak volumes about how separate we think we are from nature in life. When I go, I hope to be wrapped in a cotton shroud and placed directly into the ground, to go back to being part of the whole. No chemicals, no barriers—just skin to soil, feeding the earth as it's fed me. The way we handle death mirrors how we live—insulated from nature, filled with chemicals, trying to escape the cycle that's always happening. Of course, there are some new options out there: human composting, biodegradable urns that grow into trees, ways to return to the land more naturally. But largely, the standard practice remains one of disconnection—embalmed, encased, isolated.

This isn't how it's always been. Indigenous cultures, even older Western traditions, understood death as a return. They buried their dead in shallow graves or left them for the elements, letting the body nourish the land as it had nourished them. Today, we treat death like a problem to be solved, a mess to be cleaned up. We pump bodies full of toxins to delay decay, as if decay isn't the point. We spend thousands on caskets that defy the earth's embrace, as if we can hold onto something that's already gone. It's the ultimate expression of our arrogance—the belief that we can outmaneuver nature, even when our time is up.

But nature doesn't care about our caskets or our chemicals. Given enough time, it'll break through anyway—crack the steel, seep past the seals, reclaim what's hers. Why fight it? Why not lean into it? On the farm, when an animal dies—a sheep, a rooster, a cow—we don't seal it away. We let it go back. The compost pile hums with life because of death—bones and flesh turning into rich, black soil that grows our food. That's not gruesome; it's sacred. It's the way things are supposed to be.

A Call to Reconcile

So where does that leave us? We can't eliminate death—it's not ours to erase. But we can change how we relate to it. We can stop pretending it's the enemy and start seeing it as the partner it is. On the farm, that means culling when we must, composting what we can, feeding our soil and our people with what's left. It means raising animals with care, letting them live well, and honoring their end when it comes. It means growing food that's part of the cycle, not apart from it.

For you, it might look different. Maybe it's eating meat from a farmer you trust, not a factory. Maybe it's planting a garden and watching the bugs and worms do their work. Maybe it's choosing a shroud over a casket when your time comes, letting your body return to the earth unhindered. Maybe it's just sitting with the truth that every meal, every breath, ties you to something that's died—and being okay with that. We don't have to love death, but we have to respect it. It's the price of being alive.

The sooner we reconcile with that, the sooner we can live in a way that's real—not the sanitized, guilt-ridden half-life we've built, but a full, grounded one that honors the balance we're part of. Nature's been trying to tell us this all along—through the squirrels, the sheep, the soil, even our own bodies. It's time we listened.

CHAPTER 7

Cow vs. Solar Panel

I stand in the field with my cow, her breath puffing into the cool morning air, and I can't help but feel it: we belong here. She belongs here. The grass beneath her hooves, the soil teeming with life—it's all part of something bigger, something ancient.

Yet the world around us is obsessed with emissions and carbon, as if this element—the very stuff we're made of—is the enemy. We are made of carbon—every muscle in our bodies, every leaf on the trees, every microbe in the dirt. It's all carbon, cycling life in a dance that's been going on forever.

How did we get here, pointing fingers at the building block of existence, with headlines screaming about carbon footprints and net-zero goals, pushing wind turbines and solar panels as our salvation? The more I think about it, the less it adds up. Green energy promises a clean future yet is built on hidden costs—outsourced suffering, ravaged landscapes, mountains of waste—while my cow, chewing her cud, gets painted as a villain for her methane burps.

The green energy dream looks pristine from a distance—turbines spinning gracefully, solar panels soaking up rays, electric cars gliding silently through suburbs. But up close it's a different story—a story of exploitation and illusion. Solar panels don't sprout from the earth ready-made. Most come from China, where their production is a nightmare of human and environmental

My bull, Bubba.

cost. Silicon mining means workers choking on toxic fumes, rivers turned to sludge, and forced labor camps in places like Xinjiang. The realities are buried under the "clean energy" label, while coal-fired factories churn out these panels, spewing more carbon than they'll ever save. Wind turbines follow the same script. Their magnets need rare earths—neodymium and dysprosium—torn from places like Inner Mongolia, with mines leaving behind poisoned water and radioactive waste. The turbines themselves end up as landfill fodder after a couple decades, their blades too stubborn to break down.

Then think about the batteries powering our electric dreams. Lithium is sourced from parched South American flats, and cobalt is dug by child hands in Congo's chaos. To electrify everything we'd have to gut the planet—according to one estimate we need 400 percent more lithium, 200 percent more cobalt, and 165 percent more nickel. It just doesn't make sense.

And the carbon we're fighting isn't the monster we've made it out to be. Plants, through photosynthesis, take carbon from the atmosphere, bring it down into their leaves, sink it into their roots, exude it as carbohydrates, and feed the microbiology. Microbes feed the soil and the plants and store carbon. It's a system that's worked since life began.

The Rodale Institute says that regenerative farming could lock away more CO_2 than all our emissions combined. No mines, no slaves—just living soil,

Cows on grass doing what they're meant to do. The idea that they're harming us is ridiculous. It's obvious they are part of God's design.

living plants, and time. Sinking carbon is not hard. We don't need special technology to drive it deep into the earth; we just need to shift our agricultural practices.

Why the obsession with tech fixes, then? Perhaps it's because you can't get filthy rich building regenerative soil. The solution is simply farmers working hard to do the right thing, while feeding their families, their communities, and the world. Maybe that's why we don't see billboards or celebrities preaching agriculture. Green energy is a goldmine for the connected few.

Sometimes I can't believe I ever bought this. Standing in the field with my cow, I know she belongs here. And I belong here—we belong together. She's not a pollutant; she's a partner. The real pollution comes from factories churning out "solutions" that solve nothing. Plugging in an electric car feels good—it's quiet, smug, and modern—but where's that juice coming from? In most places it's coal or gas, humming away in some distant plant. The car is only "clean" because the smokestack is just out of sight.

In blue states we love to ban local drilling, but we don't stop the flow of gas at the pump; therefore we're outsourcing the pollution to somewhere else rather than handling it right here, where we can be responsible for it. To ban oil drilling in Los Angeles is only to pollute somewhere else, whether in the Gulf Coast of Texas or the rainforest home to the Ottawa people in Ecuador or Brazil. It's a façade, not a solution. It's like veganism's dirty secret: no meat doesn't mean no death. Plowing fields kills bugs, mice, and birds; shipping soy across oceans burns fuel. There may be zero emissions at the tailpipe, but there's a trail of wreckage upstream.

My cow's cycle is clean; an electric car's is a shell game. We don't need more tech—we need less arrogance and more scaling up of regenerative agriculture. We need more mimicking of what nature already does. Grazing cattle mimic bison herds, cover crops shield the soil, and microbes do their work. It's not sexy, but it's real, small-scale energy. Now, solar panels on a barn roof do make sense—just not a desert mega-farm. Local beats global every time.

The vegan chef who's now a cattle rancher. I wouldn't have it any other way.

We've strayed, chasing machines instead of revering the miracle of God-created carbon cycling. With my cow I feel it—we're part of this, not above it. Farmers who swap Big Ag for living soil aren't just feeding us— they're healing what we broke. It's hard work, not a get-rich scheme—that's why it's sidelined.

My cow and I will return to the earth. The wind turbines will rust in heaps. The real fix isn't coming out of some factory—it's under our feet, in the living soil.

CHAPTER 8

The Mushroom Journey and a Promise to God

In my early thirties, I was invited to my father's house in Vacaville, California, for a mushroom tea ceremony with one of the thirteen Indigenous Grandmothers. My dad, my brother, their wives, and about fifteen of us—family, close friends, and employees from my dad's restaurant, Café Gratitude—sat around a crackling fire, the air thick with sage and burning herbs as the tea settled into our bodies.

My uncle Scott was not in the circle but was working in the field, his silhouette cutting through the golden hills as the sun dipped low. I looked out across the land, watching Scott move, slightly hunched, and then something wild happened: he transformed before my eyes. His body aged fast—it curled into a fetus, then unfurled into a baby, a boy, a man—cycling through life over and over. Then a voice hit me, deep and clear: "Do you think you know better than me?" I froze. "Do you think you know better than Divine Intelligence?" No, I thought, shaking my head. I don't. The vision had just started—blood, vacuums, fetuses—not separate, but woven into a vast tapestry of life. I saw how pulling one thread, one life, shifted everything, weakening it all.

The voice came again: "You made a decision that wasn't yours to make." I'd had two abortions in my twenties. I'd brushed them off like you would a root canal or some other procedure—quick fixes, no big deal. But now I was being

Our first wedding was in the backyard. It was admittedly an awkward occasion.

shown how interconnected it all was, how those choices rippled out in ways I'd never grasped. The blood, the gore, the shifting fabric—it wouldn't stop.

"What do you want from me?" I begged. "Never do it again," the voice said. "Never support anyone in doing it. Never speak like it's a good idea. Never fund it. Mother, Father, Child—that's the Holy Trinity. That's sacred. Can you honor that?" Inside, I pushed back—I wasn't going to become pro-life, I knew—but I nodded. I could agree to that much.

The next morning, when we shared our journeys, I kept it vague: "I got some stuff from my past." How could I tell my liberal Café Gratitude crowd I'd heard God—that I'd been shown abortion was wrong, that I'd made a promise? It felt too big, too dangerous, so I swallowed it down.

A year and a half later, my dad—who'd been part of an abortion with my mom—drank ayahuasca and had the exact same vision, the same conversation, and made the same promise. When he told me, I spilled my own story, and we just sat there, stunned. It wasn't just me.

Not Chasing Feelings

Years later, after opening my Culver City restaurant, life threw me another curveball. My ex-husband—a good man, really, just going through a brutal time—stumbled in one night, drunk and unraveling. His best friend had overdosed, his brother was in prison, our divorce had wrecked him. And there, at the bar with him, was a different ex-boyfriend of mine, the one who'd helped build the place. They were trading stories, trying to outdo each other on who'd gotten the worse end of the bargain with me, their voices loud and bitter. I tried to cut my ex-husband off with a kombucha; he spat it on me. Humiliated, I hid in the kitchen behind the pizza oven, crying.

A cook found me and handed me a margarita—he didn't speak English, just made a drinking motion to encourage me—

Our second wedding—at Elias's home in Oaxaca, Mexico, dressed in traditional garb—felt much more authentic.

and I took it. Then another. One thing led to another, and we slept together. I woke up mortified, begging him, "Don't tell anyone—I'm your boss. This was wrong."

Weeks later, I found out I was pregnant. I marched into the restaurant and told him, "Break up with the host. We're getting married. I'm pregnant." He stared, speechless—he was thirteen years younger than me, undocumented, and barely spoke my language. People thought I'd lost it; some who knew my promise said, "It was just a hallucination; you don't have to do this." I wanted to believe them, but that voice echoed in me.

Two weeks later, on Thanksgiving Day, Elias and I married in my backyard, with catering leftovers, a Whole Foods turkey, and vows about duty, not love.

In January, we lost the baby.

Osmar, being folded into our family.

Friends said, "You're free now—annul it." I looked up, tears streaming, and asked God, "Why? Why make me leap, then take it away?"

Not long after this, we catered a fiftieth anniversary for an Indian family. I asked the matriarch her secret. "Arranged marriage," she said. "We built love step by step, not chasing feelings." That night, with Google Translate, my husband and I decided maybe that baby had arranged us.

We stayed together, had four kids, and fell in love for real.

The Soul I'd Discarded

Years later, a teenage boy named Osmar showed up at our farm. He'd crossed the border alone as an unaccompanied minor, feeling sick shortly after he got here, not knowing why. He googled, "Can you get sick from food in the United States?" and found Dr. Mark Hyman, which led him to regenerative agriculture, then to my YouTube channel and our website, where he signed up to volunteer.

The day he arrived, wearing a mask, I was deep in my anti-mask phase. He asked if he could come back the next weekend; I agreed, even though he had that mask on, and I let him know, "You don't have to wear that here." Rio piped up, "The air's clean out here—it's fresh on the farm." Osmar came back that next weekend and never left after that.

Eventually, I went to court and took guardianship of him. He moved into the garage loft bed first, then the guest bedroom downstairs—we called him our newest, biggest child. We were grateful for his commitment to regenerative agriculture, to the farm, to our family. One night, sitting in the TV room,

Our whole family, on the farm in California.

he looked at me and said, "The first time I saw you on YouTube, I knew I was meant to be with you, to be part of your family."

My heart stopped—I knew Osmar wasn't the exact soul of one of the children I'd discarded in my twenties. But I knew that God was giving me a second chance to care for those babies. He was mine just as much as they were.

That radical faith, trusting the path, is what's missing today. We don't trust each other, ourselves, or God, so the government steps in—telling us how to eat, raise kids, run businesses. Have we let this happen because we're disconnected from the sacred, from nature?

If we could quiet the noise and hear God's whispers, maybe we'd finally know what to do.

CHAPTER 9

Even a Seven-Year-Old Can See It

There are so many ways we're divided nowadays. For one, many people believe that race is real but that gender is just a construct.

But does this hold water in nature? Gender roles pop up across all species, tied to clear functions. Race, on the other hand—judging folks by skin color or looks—is something we humans cooked up—it's not a rule the natural world follows.

Living in Los Angeles, I was neck-deep in talks with friends and family from New York, where "gender is a construct" and "race is real" were practically commandments. I bought into a lot of it—I still do, in part. I think adults should have the freedom to present themselves however they want, with informed consent. But back then, the idea that you could pick your gender, tossing aside the body God gave you, was cheered, while race was seen as a hard line—black, white, etc.—as if skin tone locked you into one shared story.

To me, calling gender a construct is wordplay, a clever dodge around the hard fact of biological sex. Race, though? That's a construct we've built to split us apart, not to bring us together.

We've confused ourselves into silence. Modern discourse has tangled race, culture, and ethnicity into one messy, inflammatory knot—so now, nobody

My wedding party from my first marriage. All people are equally worthy in God's eyes. Why do we continue to use a manmade construct like race, which only divides us?

knows how to talk about any of it without either being called a bigot or accused of being blind to truth. But nature doesn't deal in political correctness—it deals in reality. Ethnicity is real. It's our genetic makeup, the biological roots of who we are. And that should be honored. I have one. You have one. Elias has one. We all come from somewhere.

Culture, on the other hand, is how we live, think, and behave. It's the stories we tell, the traditions we practice, the values we uphold—or don't. Some cultures promote flourishing, responsibility, and reverence for life. Others promote decay and the breakdown of the family—like much of American culture today. Pretending all cultures are equal just because all ethnicities are equally worthy is intellectually lazy and spiritually dishonest.

The political left often makes the mistake of flattening this difference—assuming that to love all people, you must love all cultures. But love tells the truth. And sometimes that truth is a critique. The political right, on the other hand, can fall into the trap of linking negative cultural behavior with ethnicity—failing to separate inherited genetics from learned behavior—and that's how you end up with repugnant ideas like racial superiority.

But race is just a cultural word. It's not a scientific one. It's a blurry amalgamation of genetics and behavior that makes it almost useless in serious conversation. If we could untangle the threads—honor the dignity of every ethnicity, critique culture where it's harmful, and stop acting like every discussion is a threat—we might just begin to heal.

Bubba Brings Babies

My seven-year-old daughter cut through the noise one day with a simple observation from the farm. She said, "When only the girl cows and heifers are in the pasture and they mount each other, no calves come. But when Bubba, our big bull, is there and mates with them, we get calves." It's so obvious a kid can see it: reproduction takes a male and a female. That's not a social construct—that's biology, plain as day. Separating gender from biological sex is a strange concept that I don't understand, but biological sex isn't up for debate. God doesn't make mistakes—you can't be born in the wrong body.

We see all kinds of behaviors on the farm. Camila, our little Highland-Dexter cross cow with horns, steps up when there's no bull around—leading the herd, protecting it, acting dominant. All the cows mount each other when one's in heat, but Camila's the one who takes charge most. She steers clear of Bubba unless she's in heat, and then they're thick as thieves, her usual standoffishness gone. Behaviors shift, sure, but her role as a cow doesn't. She calves every year, a fierce mama, her biology unchanged by how she carries herself.

We've got pigs that co-parent too—sows letting each other's piglets suckle, sharing the load like it's no big deal. Our chickens trade off sitting on eggs, one hen taking over when another needs a break. We see animals exhibiting

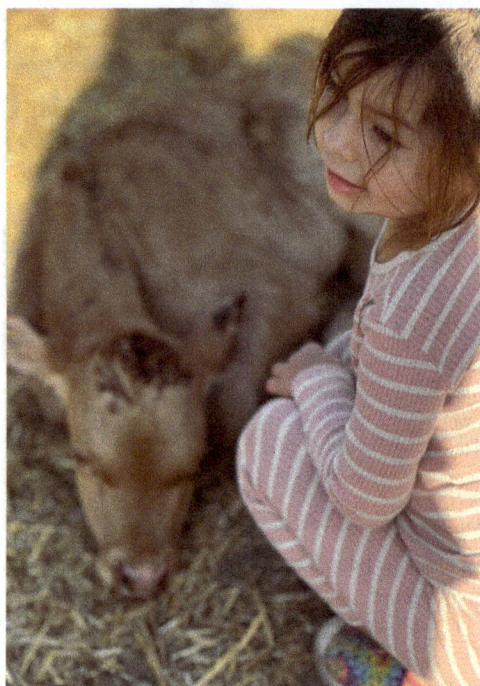

Luna with a baby calf. From a young age, my kids have seen how animals reproduce. They help during births and get to bottle-feed babies.

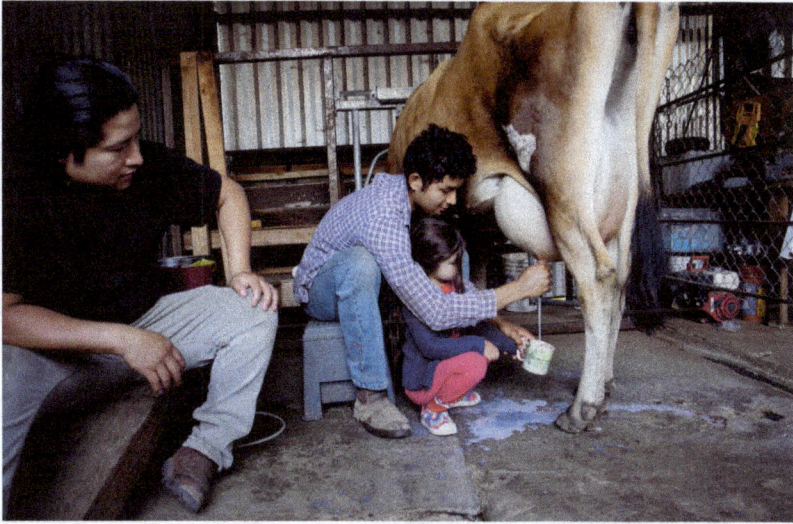

Osmar and Luna, with Elias looking on, milk Hilda, one of our favorite milk cows.

what some might consider homosexual behavior—hens pairing up, roosters getting cozy—but never once have I seen a farm animal give up its biological sex or its drive to reproduce. They don't question their bodies or their purpose. Why do we? Why are we acting like it's normal to turn our backs on what every other mammal holds sacred?

In nature, diversity isn't about division. Take our sheep: just the other morning, an all-black ewe dropped an all-white lamb. No one in the herd cares—coat color doesn't mean a thing to them. Chickens come in every shade, like our Silkies with their black skin under white feathers, prized in Chinese cooking for their look and taste. Egg colors vary too—white from hens with white earlobes, brown from those with red—but a broody hen sits on any egg in her nest, no bias, no hierarchy. Animals don't sort themselves by appearance; they're wired for survival and reproduction, not social labels. Race is our construct, devised to split us apart, while nature just keeps doing its thing.

My husband has indigenous roots deep in his bones, with skin like plenty of my LA friends. His childhood—traditional, communal—couldn't be more different from theirs, or from mine, growing up on a hippie homestead near Ithaca. I've got the same pale skin as folks from Appalachia, but my world barely overlaps with most people in West Virginia. Experience isn't skin-deep—it's culture, family, place. Lumping people by color oversimplifies everything, and it's a lie we've let fester. Meanwhile, we're obsessed with blurring biological sex, pretending it's fluid, when nature shows us it's not. Gender isn't just "a costume"—the sex of the body underneath cannot be changed by surgeries, makeup, or clothing. Every cell in that body expresses the biological sex it was born as.

Osmar helping Luna take a ride on Lola, a sweetheart who's half Jersey, half Angus.

Forsaking Fertility

What's wilder is how little we care about our fertility. Farmers get flack for ruining the planet, and while there are plenty of practices that are harmful—including herbicides and pesticides that disrupt our hormones and impact our fertility—it doesn't start or stop there. It's in every area of our life.

The real damage comes in many forms, flooding our lives—shampoo, toothpaste, deodorant, all those personal care products loaded with endocrine disruptors, leaching into our skin. We have plastics in our food containers, water bottles, and kids' toys, shedding chemicals that mess with our bodies. Our household cleaners, air fresheners, and laundry detergent poison the air we breathe and the clothes we wear. Our food is so stripped down and processed, full of crap that clogs our systems, that it's giving us COPD, asthma, and inflammation. All of this is tied to declining fertility too. We're choking on our own diet, and it's making us less able to bring new life into the world.

In addition to all this, we put abortion on a pedestal—why do we fight so hard for the right to kill our own children when family and faith should be the most important issue? Kids are taking hormones that could sterilize them for good, women are getting their breasts cut off, men are having vasectomies left and right—it's like we've got no interest as a society in reproducing. Every other mammal is driven by this—obsessed with feeding, mating, and raising their young. Why are we so busy chasing something that's "more important" than our own kids? Why are we hell-bent on finding ways to not reproduce, or committing to never doing it at all?

We're the species with science and spirituality—a relationship with God—and yet we're acting against both. Endocrine disruptors, junk food, hormones for kids—it's societal conditioning telling us fertility's disposable, that farmers and animals are the bad guys, that family's a burden on our "personal goals." We're lulled into a daze, fussing over electric cars and paper straws while global fertility tanks and our water, food, and soil get poisoned. We're sleepwalking, disconnected from nature's order.

On the farm, though, priorities are clear: water, food, safety, reproduction. Emotions don't override facts. We've got dominion—we're meant to steward this place. My pigs and chickens don't ditch their instincts to co-parent or to hatch eggs. They don't trade their biology for some trendy idea.

We've got to wake up and see what even a seven-year-old can: there's a race to connect with nature, not the machine.

CHAPTER 10

Food Is Medicine

In the United States, we swallow more pharmaceuticals than any other nation—pills for diabetes, pills for heartburn, pills for the jitters, pills for the fallout from other pills. Yet we're among the sickest people out there. Bookstores overflow with diet fixes—keto, vegan, raw food, whatever's trending—each promising to melt pounds and mend us. But most of us wind up back at square one, popping meds to hush symptoms while the real problem festers. We're hooked on band-aids, ignoring the wound.

Our food is making us sick. It's a slow drip of poison, sending us to doctors who scribble prescriptions that patch one issue and spark another. The cycle churns: a pill for the ache, a pill for the gut upset that pill triggers, a pill for the sleeplessness that follows. We're drowning in fixes that don't fix.

Nature Didn't Budge; I Did

Out on the farm, I watch my cows and goats graze on plants we'd tag as weeds—or even toxic—and they're fine. They nibble a bit of this, a bit of that, like they've got a built-in sense of what heals them. I've seen a goat munch on milkweed, which books call poison, and strut off unfazed. Maybe it's a dewormer, a cleanser, something they just know is good. They don't overdo it—except with grain. Tip a barrel of corn out, and they'll eat till they're bloated, sick, even dead. In nature, grain's not piled up like that; it's part of the plant—balanced. We humans are like cows who have access to the barrel of grain—just like a cow who eats until it gets bloat and dies, we consume refined foods— sugars, flours, processed junk—until we gorge ourselves into illness.

We're so confused. We've strayed so far from nature that we don't know what's food anymore. A bag of Skittles? I can tell that's not real—easy pass.

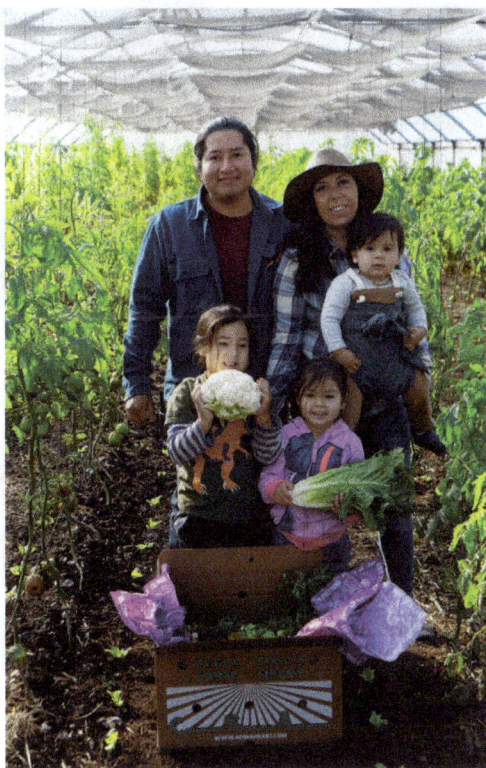

Our family in the greenhouse. It's a blessing to be able to grow a lot of our own food.

But the big stuff stumps us: meat or no meat? What's the right diet? Why are we obese? Why are autoimmune disorders, autism, and infertility climbing? What if it's all linked to what we eat? What if this diet—so detached from soil, seasons, and sanity—is denaturing us, unraveling our health?

It's not complicated. Food is medicine. Switch to whole, nutrient-dense foods—raw dairy, grassfed beef, organic vegetables, and fruits—and you can turn back almost any chronic mess. No need to be a biohacker sipping fermented kelp or chasing goji berries. Just get back to the roots, the stuff our ancestors ate without a fuss.

Weston A. Price nailed it in the 1930s. Trekking across continents to study indigenous folks—Swiss herders, Polynesian fishers, African tribes—he observed that their diets were local, seasonal, and unprocessed: fresh milk, wild game, roots dug that day. Their bodies showed it—strong jaws, sturdy frames, hardly a sniffle. Then he eyed modern folks—us—chowing on white bread,

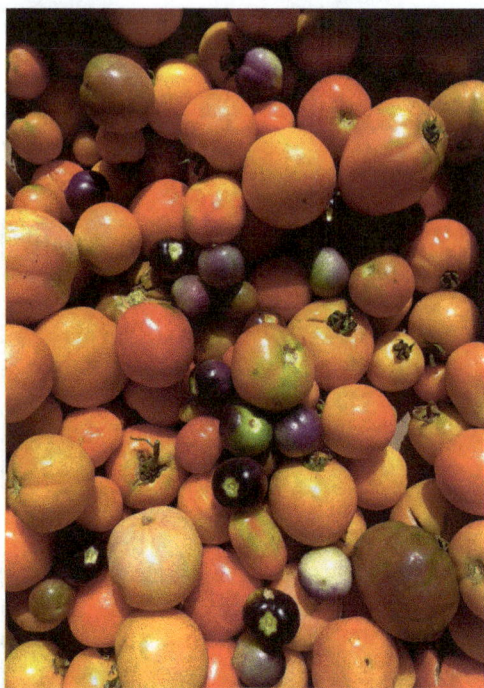

Beautiful ripe tomatoes of all types and colors—part of the whole, nutrient-dense diet we try to eat.

canned slop, sugar by the bucket. Weak bones, crooked teeth, sickness galore. Even members of those indigenous communities who'd moved to "civilization" and adopted the modern diet—those first-generation modernizers had significantly worse health conditions than their family members still living on traditional diets. We've gotten so smart that we've lost the plot. What species forgets how to eat and poisons itself into a corner?

I was sold on veganism for most of my life. Nothing could've swayed me, but nature showed me the silliness. I thought veganism was noble, but in reality, it was stubborn. Out on the farm, I watched my animals—cows chewing grass, chickens snagging bugs, goats foraging wild greens. They weren't just getting by; they were robust. For millennia, humans ate nutrient-dense foods—meat, dairy, eggs, plants straight from the earth. Now we strip the good out—refined grains, fake fats, soy glop—and wonder why we're crumbling. Nature didn't budge; I did.

Look at leafcutter ants. They're tough as iron. People think they eat the leaves they cut, but no—they tote them underground to grow a fungus. That's their only food. They don't import, don't outsource, and don't eat what they don't cultivate. It's why they're nearly indestructible—local, self-contained, no weak links.

If ants can master that, why can't we? We're trucking in apples from halfway around the world, beef from continents away, greens from who-knows-where, while our own land sits quiet. We've handed our lifeline to far-off factories and cheap labor, then we act stunned when it's tainted or gone.

Grow, Swap, Buy Local

We need to snap out of it. Local food hubs aren't a pipe dream—they're survival. Picture it: gardens in every yard, farms ringing every town, dairies a bike ride away. No one could choke us with poison from a distance if we fed ourselves. An ant gets it, but we balk. We don't want to sweat. We'd rather snag a frozen pizza than break soil. Convenience is our crutch, and it's snapping our spines.

The eggs from our chickens always amaze us with their various colors.

This isn't just about health—it's national security. We obsess over local energy and local factories, but local food? Silence. In a pinch, it's air, water, food—that's the list. We've got to keep our water pure, our air clean, and our food growing nearby. One storm, one war, one broken chain, and those shelves empty fast. Then what? Pills won't fill your belly.

Food is medicine, and we wouldn't need half the drugs we take if we ate seasonal, regional, single-ingredient stuff. A potato from your patch, a steak from your neighbor's herd, milk still warm from a cow—that's it. We're addicted to processed trash—sugary, salty, boxed-up junk that'd baffle our great-grandparents. My kids don't know what a Twinkie is, and I'm glad. They eat what we raise: eggs from our hens, greens from our rows, beef from our pasture. They're sturdy, sharp, and rarely do down with a cold.

I've seen this play out. One spring, our cow Una started limping—stiff joints, slow gait. The vet pushed meds, but I waited. We turned her loose in the back pasture, thick with wild herbs. A week later, she was trotting, chew-

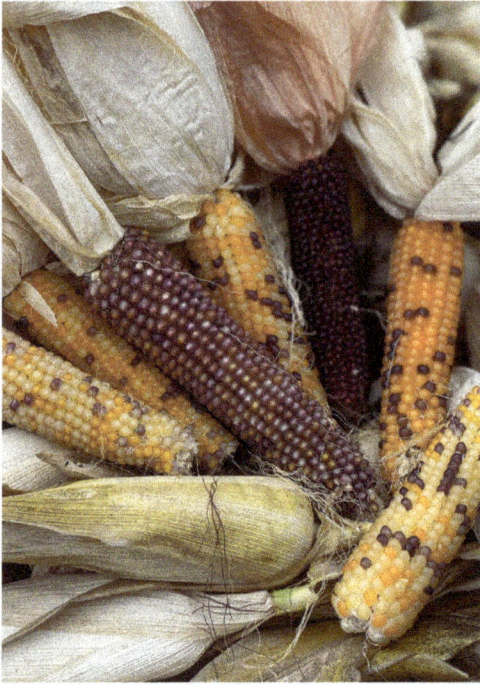
Popcorn—grown right here.

ing on what she needed in the plants we call weeds. She knew.

Last year, my kids had a cough that lingered. The doctor waved antibiotics, but I passed. We leaned on broth—bones from our steer, simmered with garlic and thyme from the garden—and raw milk, straight from a cow. Three days and they were back tearing around the yard. Food did that, not a prescription.

It's time to reconnect—with nature, with our food, with each other. Start small: herbs on your stoop, a market haul, a meal with no factory fingerprints. Ask where it's from. Grow something—anything. Link up with neighbors; swap what you've got. Our health is on the line, and we are only as food secure as our neighbors are.

We must start somewhere—plant a garden or a small orchard, get a flock of chickens, have a couple of milk goats, and trade with your neighbors. We must start depending on local food. We must start creating local food. This is a national security imperative. If leafcutter ants can figure it out, we can as well.

CHAPTER 11

Discomfort Is Not Dangerous

In today's society, discomfort is treated like a plague—something to dodge, medicate, or banish. We've convinced ourselves that unease is a danger signal, a threat to our peace, rather than what it really is: a spark for growth, a rite of passage, a sacred call to rise. This comfort fixation has wormed its way into everything—parenting, work, relationships—and it's left us fragile, cracking under the lightest load. We've forgotten how to see discomfort as a guide, not a foe.

I learned this early. My parents didn't shelter me from life's sharp corners. If I came home upset—say, I got in a fight with a friend or did poorly in class—they'd listen, but they wouldn't rush to fix it. Instead, they'd ask, "How did you participate in this?" It wasn't about fault; it was about ownership. If I'd argued with a friend, they'd nudge me to own my words. If I'd stumbled in school, they'd ask what I could've done differently. No hand-holding, just empowerment. It sank in deep: discomfort wasn't something to run from—it was something to face, a chance to toughen up. That shaped me more than I knew.

Meeting the Job's Demands

When I owned restaurants in California, I carried that same mindset

Hard at work in the kitchen.

into my management. My style echoed my parents' leadership: personal accountability was non-negotiable. I wasn't there to baby my staff or shield them from the job's demands—I was there to help them meet it head-on.

But I quickly saw how that rubbed against a culture that flinches at discomfort. One night, during a hectic dinner rush, a young waitress—early twenties—came to me, tears pooling. "Can you take this table?" she asked, voice shaky. "They're being rude." I stopped mid stride—checking a ticket with the expediter—and turned to her, concerned. "Did they say something wrong? Were they harassing you?" I needed to know if this was real trouble. She hesitated, then shook her head. "No, I messed up their order—forgot their appetizer—and now they're frustrated."

I saw it plainly: this was not a problem but a possibility. "This is your moment," I told her. "Go back out there. Apologize—mean it. Make it right. You've got this." She looked at me like I'd asked her to tame a lion. The next day, before her shift, she pulled me aside. "I'm quitting," she said, her tone clipped. "You didn't have my back last night. I didn't feel safe."

Safe. That word stuck like a splinter. She wasn't unsafe—not physically, not emotionally in any true sense. She wanted a buffer from her own slip-up, from the discomfort of facing annoyed customers and fixing it. I hadn't given her that buffer, so I'd let her down, in her view. We've raised folks to think discomfort is harm, accountability is an attack, and mistakes should come with no cost. When that bubble pops, they're gone.

It wasn't a one-off. Another server, Jess, quit a few months later, and her story sealed the lesson. Jess was sharp and quick with a smile, but she coasted—she was late to shifts, dropped orders, and took half-measures. One night, I'd had it, and I pulled her aside after service. "Jess," I said, keeping it straight, "you've been late three times this week, and last night you forgot two orders. The kitchen was scrambling, and the guests were upset. We need you to step up." She nodded and promised to try, and I thought we'd turned a page.

A week later, she told me she was done. "Why?" I asked, wanting the real answer. She shrugged. "The patriarchy," she said, smirking like it was a quip. I dug in. "Did someone say something? Do something off?" I was ready to fix any real problem. "No," she said, "it's Michael. He's always bothering me—asking me to run drinks, telling me to take the bus tub back, pointing out trash on the floor I walked past and didn't see, asking if I can check for plates to run. It's like he doesn't even want me to rest or relax for a minute. He wants me to be working. It feels harassing. Every moment, he's on me to do something for the restaurant."

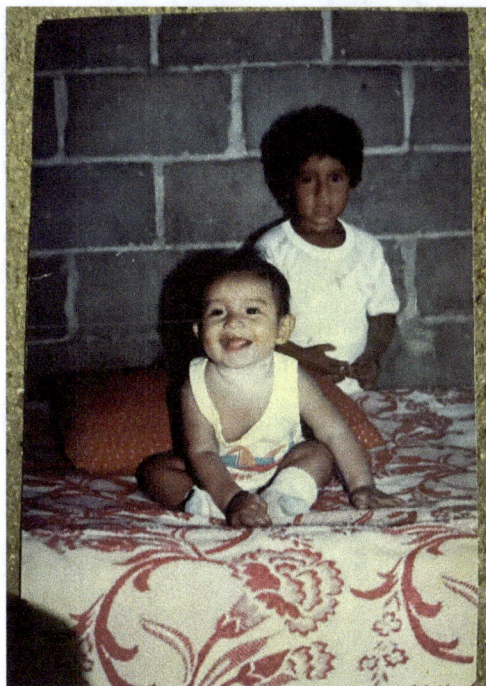

Elias and his younger brother, Jose. They grew up in Mexico with three block walls and a thatch roof, without running water or electricity.

I couldn't help but grin—not at her, but at the gap. "That's his job," I said. "Michael's the manager. He's supposed to keep things rolling. And yeah, he wants you to work—because that's what I want too. That's what keeps this place humming." She didn't respond, just looked away. To Jess, Michael's expectations—clearing tables, running food, keeping it tight—felt like oppression. Not because he was harsh, but because he wouldn't let her slack. He held her accountable, and that discomfort was too much. She tagged it "the patriarchy" because it was easier than owning her reluctance to work. She was out within days.

Discomfort Is How We Grow

Running a restaurant isn't a breeze. Orders pile up, customers grumble, equipment breaks. You don't get to check out when it's hard. That's the gift: the hard things are what sharpen you. My parents knew that, and I tried to instill it. Most didn't bite—they'd been taught discomfort is a cue to retreat, not rise.

Real life doesn't coddle either. Farm life is blunt. The cows don't care if you're spent—they need feed. The beans don't wait for your mood—they need water. Parenting's no different. I've breastfed for ten years straight, my body a lifeline for my kids. I've changed diapers daily for a decade—thousands. I've been pregnant six times in twelve years: four births that remade me, two losses that tore me open. There's no pause, no "self-discovery" jaunt to Bali. It's relentless, holy service—raw, draining, radiant. I wouldn't trade it.

This is where "my body, my choice" starts to unravel for me. It's a battle cry that sounds strong, but it doesn't fit motherhood's truth. Motherhood isn't about choice as we've sold it—it's about surrender. Your body becomes a nexus between the divine and this physical world, a bridge where miracles like childbirth unfold. Bringing life forth is gritty, messy, and incredibly uncomfortable—hours of wrenching pain, exhaustion that sinks into your bones, a body pushed past its edge. Yet it's one of the most beautiful moments of existence. That's the paradox we've buried: the sacred often comes through struggle. Your body isn't just yours—it's your baby's first home, your husband's refuge, your toddler's comfort. You give your energy, your sleep, your being—not

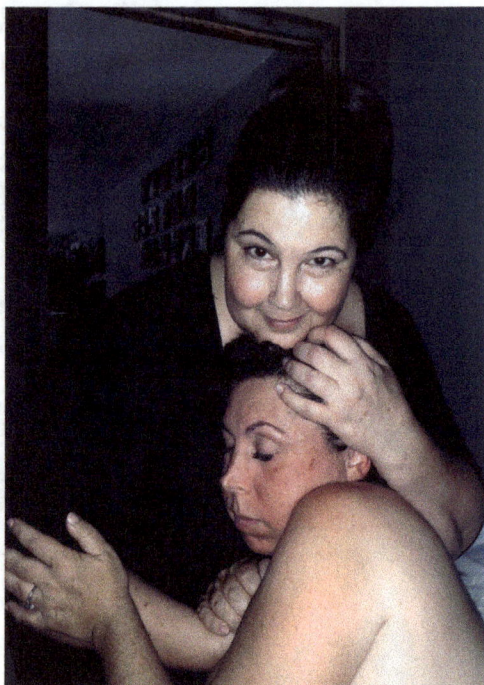

Giving birth to Sol at home, at age 42, with my midwife, Robyn. Giving birth is uncomfortable—but it's one of the most beautiful moments of existence.

because you "chose" it like picking a dessert, but because you've surrendered to something greater.

We've conditioned young women to see their bodies as tools for personal gain—sex, power, looks. But nothing prepared me for the gorgeous, total annihilation of self that motherhood demands. It's not easy—nights of tears, mornings of doubt—but it's the richest reward I've known. We need to train our girls for that, to be comfortable with discomfort, because those

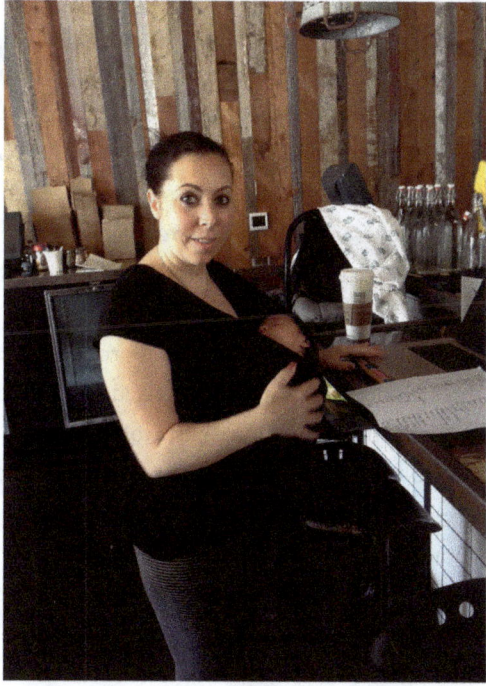

At work with a baby strapped to my chest—something I did practically continuously for years.

holy moments—birth, nurturing, sacrifice—demand it. I want my daughters—my two-year-old with her muddy boots and my seven-year-old with her sharp mind—to get this. Their bodies are sacred, not just because they're theirs, but because they tie them to something bigger: their families, their babies, their communities, their husbands. That's not oppression—it's a gift, a sacrifice, a blessing all at once.

Discomfort is how we grow. Trauma's real, but it doesn't have to rule us. Animals show us—after a predator scare, a goat will shake, shedding the fear, then graze like it's nothing. We humans cling. We replay the hurt, weaving stories around it. I've had trauma—sexual assault in my music industry days. Men with money and power wanted access to my body, and so they took it by force. It was wrong. It cut deep. It left marks. But I chose not to carry it forever. I understand what it is to have compassion for young men filled with testosterone, lacking wisdom, trying to find their way in the world. I did dumb things when I was young as well. It doesn't take away what they did, but it gives me the power to shake it off. I felt it, faced it, let it go—not because it was small, but because I wouldn't let it define me.

I know folks wrecked by less. A hand grazing their bottom at a bar, a crude remark—they've turned those into lifelong weights. They carry more trauma from that than I do from being violated. Why? They were never taught to sit with discomfort, to process and release. They were told every unease is a wound, every slight a scar. I'm not dismissing their pain—their pain is real. But I'm reclaiming our power. What happens to us isn't who we are. How we respond is.

My husband embodies this. Born into discomfort—dirt floors, no power, three cinderblock walls—he walked to America, boots hitting dirt. There are no complaints from him—just work. As a kid, he'd lie in a hammock under a mango tree, dreaming: a big truck with dual rear wheels, a wife he loved, fruit trees, kids, a farm, being the boss without the details. It all came true—part miracle, part grit, all by the grace of God.

Then one day, he said, "I think I'm depressed." It jolted me—he's not that guy. "Everything I dreamed of in that hammock came true," he said. "The truck, the farm, you, the kids, the trees. And I still feel the same." I told him the truth: outer wins don't heal the inside. Gratitude does. Faith does.

He shook it off fast—he's a survivor, forged in fiercer flames than most. But it struck me: he looks at Americans, born into abundance, and can't fathom our fragility. He sees possibility. We see barriers. He knows hunger—tortillas, lard, and salt for days. So he savors our farm's plenty in a way I sometimes forget. Discomfort taught him that. It's a lesson we're losing.

We've all heard it: hard times create strong men, strong men create good times, good times create weak men, and weak men create hard times. How do we break that cycle? We choose discomfort—on purpose, as holy. We teach our kids to face it. Farming does that. Parenting does that. Life does that, if we let it.

The pandemic proved it. Pregnant, breastfeeding, unvaccinated—I've never had a shot—some employees reported me to OSHA and the health department. Not because I was sick, not because they were hurt, but because my choice unsettled them. We've built a world that demands protection from feelings. But life doesn't yield. Birth, marriage, purpose—they're uncomfortable. And they're the richest parts of being alive.

Discomfort isn't dangerous. It's holy. It's the fire that forges us. I want my kids—resilient, grounded, fierce—to rise when life calls, not run.

CHAPTER 12

We Belong Here

E mbracing traditional ways of living—hands in the dirt, food from the field, life close to nature—led me to a truth that hit like a thunderclap: we belong here. Not as intruders, not as a blight, but as part of this earth, woven into its fabric.

Yet there's this creeping idea out there, whispered in classrooms and shouted through policies, that it's fine for us to be poisoned—that it's no big deal if male sperm counts have crashed 50 percent in forty years or if fertility struggles shadow every developed nation. The modern green movement often runs on a quiet engine: humans are the problem, a species to be culled or contained. They don't say it outright—not usually—but the vibe seeps through: we're a plague, a stain on the planet, better off walled away from nature so it can limp back to health without us.

That's the ugliest lie I've ever heard.

We belong here. We're not separate from the whole—we're in it, of it. Sure, we've messed up. We've scarred the soil, fouled the water, choked the air—I won't pretend otherwise. But the fix isn't to shrink back, to abandon nature like it's a museum exhibit we're banned from touching. The real power, the truest union, comes when we step into nature with respect, with understanding—using our brains, our hands, our science to amplify the beauty in God's design.

We're not helpless—we're potent, brimming with potential. But when we buy the lie that we're a plague, that there's too many of us, we crumple. Fear takes over, apathy sets in, and we stop doing what's right. We let the poison flow unchecked.

Bloating Ourselves

I used to feel that apathy myself. Before I grasped regenerative agriculture, I'd look at the world's mess—dead soil, sick rivers, junk food everywhere—and think, "What's the point?" It felt too big, too broken. Then I learned there's a solution, simple and strong, right under our boots: work with the land, not against it. That flipped a switch. I went from numb to obsessed, hungry to weave myself into nature's web, not stand outside it.

I've watched this unfold on our farm. Take the pasture we let go wild one season—knee-high with grasses and herbs we'd never planted. The goats wandered in, nibbling here and there, and the next year that patch exploded with life: richer soil, thicker growth, bugs and birds we hadn't seen before. We didn't abandon it; we joined it, guided it, and it sang back.

We also learned, the hard way, how close we are to nature's limits. One year we took a trip to Mexico, leaving our two goats, Ely and Oolu, with my uncle. He didn't feed them grain—just pasture, the way it should be. My husband, though, had been giving them a steady dose before we left. When we got back, Oolu was dead, bloated stiff in the pen, and Ely was teetering—swollen, gasping. We called the vet fast, pumped her with fluids, and walked her till she burped it out. She pulled through, barely. Grain is the easiest way

Nature explodes with life.

to kill a ruminant—too much, too fast, and their guts can't handle it. In nature they'd never find a barrel of it; it's us who dump it on them. We're kind of like those goats, eating ourselves into ruin with refined junk—sugar, white flour, processed trash—gorging until we bloat with disease.

Where did we get this idea that we don't belong? That we're the enemy? Why do we swallow it whole? Why, as a society, do we shrug at being poisoned? We're numb to the flood of herbicides—glyphosate soaking our wheat, pesticides coating our apples, fluoride in our taps, endocrine disruptors in our plastics. Then there's the sky: barium, aluminum, strontium raining down from planes under vague "environmental" banners. I've stood in my field, kids at my side, watching those streaks spread out, and thought, "This isn't right." We've told the world—governments, companies, whoever's listening—enough is enough. Stop poisoning us. We belong here, and we're not going quietly.

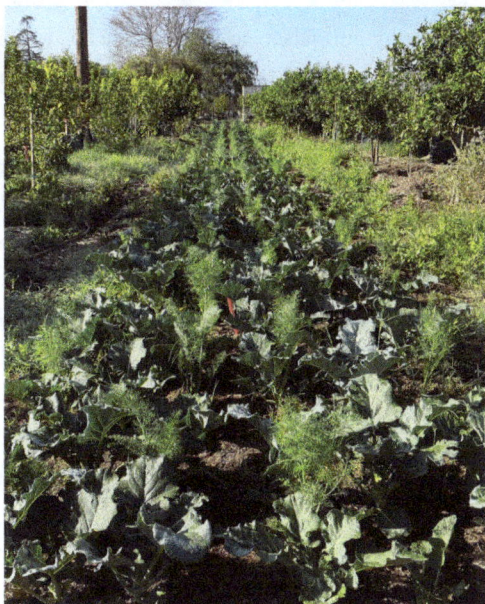

I used to feel helpless in the face of the world's messes—dead soil, junk food, pollution. The solution is to work with the land instead of against it.

Prizing the Essentials

Agriculture threads through the Bible like a lifeline, both literal and deep with meaning. Genesis 2:15 lays it out: "The Lord God took the man and put him in the Garden of Eden to work it and take care of it." Right from the start, we're tasked with tending creation—not trashing it, not ignoring it, but stewarding it.

Leviticus 25 talks about letting the land rest every seventh year as a sabbath for the soil. Deuteronomy 22:10 pairs oxen and donkeys differently, hinting at balance in how we work the earth. After harvest, scripture nudges us to bring animals in to glean the fields, fertilizing naturally.

These aren't just old rules; they're rhythms regenerative agriculture echoes today. On our farm, we rest patches of land, rotate cows and goats through it, and plant cover crops like rye and clover to heal the dirt. It's us taking our place, partnering with Mother Nature. The world wouldn't be better without us—it'd be emptier, missing a piece.

Regenerative farming isn't some crunchy fad—it's potent. Integrate animals—cows dropping manure, chickens scratching it in. Plant cover crops to hold the soil, mix perennials and annuals for depth, ditch the chemicals that kill the good with the bad. You're not just growing food; you're stewarding the soil food web, co-creating with God.

The proof's in the soil. When we started our twenty-acre farm in Southern California, it was rough—dry, thin soil, yields that barely broke even. Five years in, it was a different story. We pulled $500,000 worth of produce off that land one season—berries, greens, roots—more than we could've dreamed. That's what happens when you work with the land, with God, not against it. I remember walking the rows that fifth spring, my two-year-old tugging at my pant leg, dirt under her nails from "helping." The air smelled alive—sweet from the strawberries, sharp from the herbs. That wasn't luck; it was harmony.

We can't out-science God. Tools? Sure—tractors, soil tests, weather apps—they're gifts to sharpen what's already perfect. But this arrogance—thinking we can engineer our way past divine design—is a road to ruin. Look at industrial ag: monocrops drenched in Roundup, soil turned to dust, rivers green with runoff. That's us playing God and losing. Or take seedless grapes—convenient, sure, but sterile. A dead end. Nature's got seeds for a reason; they're life's pulse. We've got to stop tweaking creation like we know better and start enhancing it like we belong to it.

Why doesn't society prize the essentials—reproduction, food, and water? They're shoved low down the list. Kids? Too many see them as roadblocks to the good life—travel, careers, freedom. I've been pregnant six times in twelve years—four births, two losses—and every day's a reminder: children aren't burdens; they're roots.

Food? We barely think past the grocery aisle or DoorDash ping. I've had folks ask, "Why grow it when you can buy it?"—like food's just fuel, not medicine, not connection. Convenience is our church, and we are devout. My kids have never eaten at McDonald's, and I think that's okay. We still struggle with

Children aren't burdens. They're roots.

fussiness, but they eat eggs from our chickens, tortillas made from our own corn, beef from our own herd, and vegetables that we grow. When my kids get sick, I don't run to the doctor. I don't reach for antibiotics. I make them bone broth, elderberry syrup with local honey and our own elderberries, a spilanthes tincture that we grow here on the land, mullein, and nutrient-dense food, and they always get better.

Water? We turn the tap and forget it's a miracle—until it's brown or gone. On the farm, my seven-year-old hauls buckets from the well when the pump lags, grinning like it's a game. She gets what we've lost: these things matter. Grow something—anything. There's something extraordinary about being in that relationship with nature and God's creation, planting a seed and bringing forth food. Even a little meat rabbit setup in your backyard works. I've got a friend, Brandy—when I met her, she told me they were raising meat rabbits in their suburban lot. Her daughters helped harvest them, skinning and all. Back in my vegan days, I was appalled—my jaw hit the floor. Now? I admire her. Eggs and meat from her backyard, raising kids who know where their food comes from, who aren't afraid to get their hands bloody—that's real.

Farming's my church. Hands in the soil, gut humming with microbes from raw milk and homegrown veg, standing on land we've nursed back to life—that's where I've felt God closest. No pews, no stained glass, just the wind through the oaks and the hum of bees in the lavender.

CHAPTER 13

The Priority of Life

*A*nimals—whether they're out in my pasture or roaming wild—put reproduction first. It's everything to them.

I've seen a cow drop a calf in a howling rain, mud caking her legs, and she just gets to work, licking it clean like it's another day. Goats push out kids even when they're staggering with exhaustion, bleating through the ache like it's nothing. Hens hunker down on eggs till they're gaunt, feathers a mess, just to bring those chicks into the world.

It's raw instinct—they don't hem and haw or wait for the perfect setup. Us humans, though? We're a whole different mess. We shove kids to the bottom of the list, chasing jobs, freedom, whatever we've tricked ourselves into thinking matters more. We juggle calendars, debating if kids fit our big plans. Nature doesn't waste time on that, though.

I was almost thirty-seven when my first kid came. There was a real shot I'd never have any—I was too deep in my own grind to care. Society slapped "successful" on me—running restaurants, working seven days straight, doing whatever I damn well pleased. I didn't think I was missing out, mostly because I hadn't a clue what I was missing. Everyone said wait—build the career, stack the cash—kids can come later. I swallowed it whole. When a doctor tagged my first pregnancy "geriatric," I rolled my eyes and fired back, "Or financially stable," like I'd outsmarted the system.

I was smug as hell until we lost that baby. That hit me like a sledgehammer. I'd be in the restaurant, plating up veggies, and spot a pregnant woman shuffle in—belly full, eyes bright—and this empty ache would rip through me. My two sisters-in-law were pregnant when I should've been, their bodies blooming while mine stayed still. I'd plaster on a grin, hand over their plates, and feel the hole widen. I'd always figured kids weren't some grand thing. I was dead wrong.

Very pregnant, with Sol.

We got pregnant again quickly. Nine months later, I was in a fight I never saw coming—three days of labor at home, my space, my rules. I'd pictured it soft: candles low, tub warm, my midwife Robyn Pool right there. I hit 7 cm dilated, thinking I was close, when something shifted. At home, every contraction—coming every three minutes—felt like a knife in my pelvis, stabbing my anus so hard my body shut down, reversing from 7 cm to nothing.

Robyn's voice broke through, calm but firm: "You're exhausted, and you're going backward. I don't know why, but we need the hospital." I fought her—no way was I stepping into that medical grinder. I could do this. We're built for it. But my body wasn't listening, and neither was she. So we went.

The shift was a gut punch—from my cozy tub to a stark room split by a curtain, some stranger groaning next door. Contractions slammed my body every two minutes, so only one person could come in. My mom gripped me tight as the doctor fumbled, jabbing the epidural into my spine, missing the spot over and over while I shook like a leaf—sweat dripping, teeth clenched, the room spinning with every botched try. Once it took, the pain dulled, and I could relax—my body unclenched and started opening again. This was so distant from the home birth I'd imagined.

Doctors pushed for a C-section—three twelve-hour shifts of them, each one a new face with the same song: "We need to cut." The first was a stiff guy in glasses, clipboard in hand, rattling off risks like a robot. The second, a woman with a tight bun, got snippy when I said no, muttering about liability as she stormed out. Near midnight, the third—a gray-haired man—leaned in close, voice low, "You're risking it all," like I hadn't heard it twice already. I said no every time, digging in my heels, jaw set. My husband, just 23, was a wreck—pale, pacing, slipping out for air whenever it got too thick, his boots echoing down the hall. My mom sat there, cool as ice, scrolling sweaters on-line while I drifted in and out, half-gone, the beep of monitors a dull hum in my skull. A nurse buzzed around—perky one second, snippy the next, then perky again after a long bathroom break that stank of self-soothing, her perfume trailing back too sweet. They turned off the epidural to push, and that stabbing roared back—four hours, no meds, pure hell.

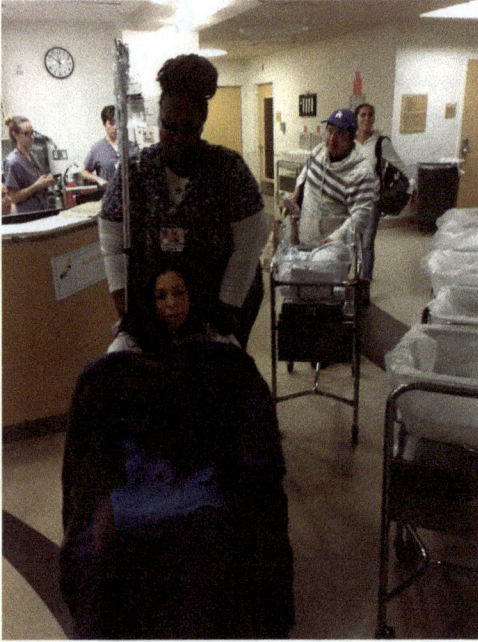
Rio's hospital birth was traumatic.

Finally, they saw it: my tailbone, busted in a car wreck at 16 and healed all wrong, was blocking the baby's head and was getting slammed with every contraction. That's why I'd reversed at home—my body was shielding itself. I was beyond spent, but they missed it in the notes. The doctor reached in and snapped my coccyx—no epidural, no heads-up. I screamed, pain blazing through me, a white-hot jolt that echoed in my bones, but then—relief. That break stopped the stabbing and cleared the path. It hurt like fury, but it was better than the relentless jabbing I'd endured.

Then meconium hit. The NICU team barged in, surgeons on their heels—over twenty people in that cramped room, voices overlapping, machines beep-ing wild. "Mollie," the doctor said, sharp, "we need him out now. Vacuum, three pulls. More than three pops, it's brain damage risk. If you can't, I'll push him back for a C-section." Push him back? That was a nightmare I couldn't face. I was drained—days of labor, broken bones—but I dug deep, clawing into some reserve I didn't know I had. Episiotomy, vacuum on, "Let's go." One heroic push, every muscle screaming, and Rio was out.

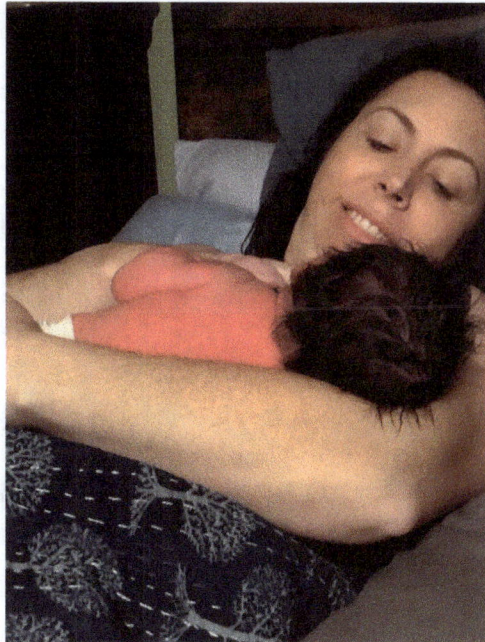

Moments after giving birth at home to Luna—a much smoother experience.

The room exploded—cheers from the NICU crew, surgeons, my family, like we'd won a title, their claps bouncing off the sterile walls.

I wasn't cheering. I was shattered—hurt, hollow, craving sleep. No love rush, no oxytocin wave, just damage and fatigue, my body a shell. They wheeled me off, nurses hovering, barking at me not to doze with Rio on my chest, their voices sharp over the hum of fluorescent lights.

The following days were all a haze. I couldn't leave until I pooped—something that was terrifying with that cut, every twinge a reminder. Migraines crashed in—the epidural's parting shot—and they told me to take caffeine every thirty minutes. Gray, breastfeeding, chugging coffee with a splitting head—I looked at him, knowing my duty. But that big love? It was missing. Was I broken? Too solo too long? No—it was the trauma. Later, I'd see it on the farm: a goat after a brutal birth, dazed, turning from her kid, eyes blank. It happens.

Luna, my second, was night and day. Labor hit after a full restaurant shift, midnight creeping in, the kitchen's clatter still ringing in my ears. My best friend Mimi was deep in cancer—chemo eating her, barely mobile. I didn't

Pregnant with Sol—with Rio and Luna at Sow a Heart Farm.

want to wake her unless it was real. But it was—fast, fierce, a freight train I couldn't slow. Robyn got there as I clung to the shower, steam my shield, water pounding my back like a heartbeat. "Get out," she said. "I'm not getting soaked if we've got hours."

I crawled to the bed, legs trembling; she checked. "Just in time. Baby's coming." She asked my husband to catch, and he nodded, hands shaky but ready, sweat beading on his brow. The door creaked—Mimi, wheeled in by her mom, frail but fierce, her bald head catching the dim light. Robyn blocked it, guarding the space, but I said, "She's good." Mimi lay beside me, eyes shining despite the hollows. "I see the baby," she said, then grinned weakly. "She's got more hair than me." I laughed, a ragged sound. "Everyone does right now." "True," she chuckled, her voice thin but warm, "but she's got a full head." All mine did.

Luna came smoothly—no chaos, just us, her first cry soft in the quiet. I showered, slid into my own sheets, held her skin-to-skin, drifting on my terms, the world outside fading to a hum. Worlds different from Rio.

Sol, my third, hit on the farm in Fillmore. I woke at 4 a.m., labor humming softly in the dark, just me and the stillness, the house creaking faintly. It rolled steady till the farm woke—workers hauling crates, walkie-talkies crackling, sun spilling through the cracks. The bustle flooded my space—boots thumping, voices calling, the day's grind kicking in—and labor stalled cold.

I felt it—my body pulling back, overwhelmed by the noise. "Turn off the radios," I told my husband, voice sharp. "Cover the windows tight. I need the dark." He did, piling blankets high, and the cave worked—labor roared back, a tide I could ride, and soon Sol was in my arms, warm and loud, right there in my room,

Just after Sol's home birth.

his cries cutting through the muffled quiet. Mimi wasn't there—her first miss—but her mom stood in, steady as a rock, her hands clasped tight. We lit a candle by Mimi's picture, held her in our hearts, and I felt her spirit anyway, a quiet presence in the shadows.

Estrella, my youngest, came at 45. Doubt gnawed—knees creaking, body groaning under the years. But I told myself: I'm strong enough. I'm built for this. Labor started gently; I didn't wake anyone, just sat on the toilet, breathing through it while the house slept, the cool tile steadying me. I ran a bath, sank in, and wept—not for pain, but for all the women that would never experience giving birth being connected to God, fully awake, fully conscious, fully comfortable in their own space. I prayed, thanking God for this experience, awake and conscious, unplugged in my own space, bringing life forth—the greatest gift. It's interesting—the most profound moment of a woman's life, most often they are sedated.

I texted Robyn that I was in labor but fine; she drove up, parked outside, the engine off. Hours later, I called her in. She checked: "Barely dilated. It'll be a while." I knew better, feeling it deep. Minutes later, I stood—one push, and Estrella slid out, the cleanest birth ever, everything but her flushing away.

Robyn, ungloved, gaped, her hands frozen mid-air. My husband woke, texted Osmar, our foster son, and roused the kids and my mom: "Your new baby sister's here. Come upstairs." We piled on my bed—four kids, my husband, Osmar, my mom—staring at the miracle of Estrella, this tiny life pulsing against me, her breath a whisper on my skin.

Through these births—hospital grind and home's grace—I've learned that intervention is a razor's edge. Less is better, but some saves lives. Knowing when is what requires wisdom.

On the farm, a rough birth can break a mom's bond. We used to pull the kid and bottle-feed it, but those babies grew up shaky, ditching their own later. Now we hold the mom and nurse the kid every few hours. Instinct kicks in, and love grows. It's better.

Writing this, one kid on my lap, another at my breast, it's clear: no business snuggles to sleep. No restaurant hears a bedtime story. A growth plan doesn't share dinner. A P&L won't fill your arms. I chased that once—work, wins, the grind—thinking it was enough. It was hollow.

I'm grateful God gave me these kids, late as I started, their sticky hands and loud laughs the realest thing I've got. Don't wait, ladies.

CHAPTER 14

The Forgotten Instincts of Parenthood

*P*arents all want to do the right thing. But in today's world, even amid the countless experts, books, and systems dictating how to raise children, we've lost touch with the most fundamental truths of being mammals. We've overcomplicated something that should be guided by instinct, not industry.

Take, for example, the common practice of putting a baby in a separate crib—sometimes even in another room. This is completely unnatural. Nowhere in nature do mothers sleep apart from their young. You won't find a lioness sleeping in a different den, a bear in a separate cave, or a wolf in a separate part of the forest from her babies. Mammals sleep skin to skin (or fur to fur) with their offspring, responding to their needs immediately, as evolution designed.

I didn't start out thinking I would be one of those co-sleeping parents. In the hospital after my first traumatic birth, I was exhausted. I just wanted to sleep, so keeping my baby on my chest and nursing felt like the only way to get rest. But the nurse kept coming in, telling me I wasn't allowed to sleep with my baby on my chest—that he had to go into the plastic roller cart next to my bed. I remember feeling exhausted and frustrated.

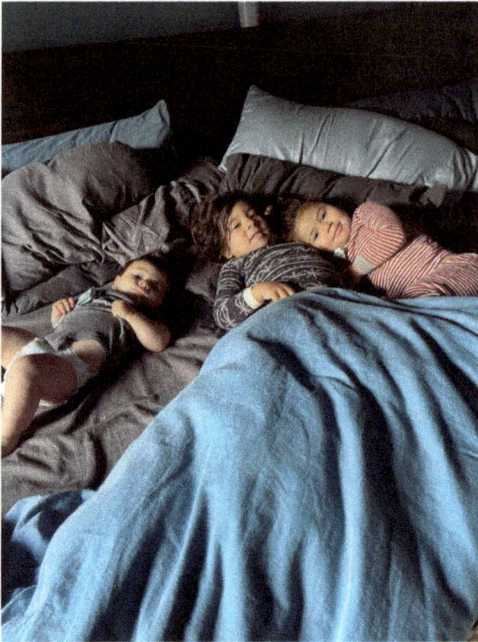
The kids in the family bed.

Once we were home, I kept Rio in the bed with me out of sheer necessity. I was still recovering, and getting up, fully awake, every time he needed me seemed absurd. People warned me, "If you start them in your bed, you'll never get them out." But once we were in that routine, I couldn't imagine doing it any other way. And when I became pregnant with my second while still nursing my first, it was natural to bring Luna into the same bed. By the time we were expecting our third, we made a custom bed—two California King mattresses pushed together. Plenty of room for everyone.

We've largely left it up to our children when they want to transition to their own rooms. I have never spent a night away from them since the day my first was born. He's ten now. Has it cost me financially? Absolutely. There have been speaking engagements and lucrative opportunities I've turned down. Podcasts that only record in person, events that required me to travel—I've said no to a lot. But if my kids know one thing, it's that I'll be there when they go to bed, I'll be there when they wake up, and I'll be there if they need soothing in the night. I am by no means a perfect mother—I work too much, I get distracted, I lose my patience—but I try to hang my hat on this one unwavering truth: my presence.

Part of why we set things up the way we later would when we moved to Texas—the restaurant, the Airbnb, the event space, all on the property—is because I spent too many years having to decide whether to leave the kids at home or take them with me. Early on, that meant bringing them to festivals to sell food, hauling our first two kids along in an RV parked close by so I could keep breastfeeding them. For years, almost every weekend of festival

Rio on our newly built bed—two California king mattresses pushed together.

season, we were at a different event, working 15 to 20 hours a day—cooking, serving, hustling—with Rio and Luna right there. I'd duck into the RV between shifts to nurse them, their little bodies pressed against me amidst the chaos of grease and crowds. It was exhausting, but it kept us together.

Later there were years with a baby strapped to my chest, running food in a restaurant, balancing plates while nursing through the chaos. Then, when they could go several hours without breastfeeding, I'd leave them at home, but that never felt right either. This setup isn't as good as if I could just be a stay-at-home mom and give them my undivided attention all the time—for sure.

One day, Luna came up to me and said, "Mom, I seen on TV there's moms with no job. I'm gonna be your mom with no job when I grow up, so my kids don't have to miss me." That broke my heart and made me want to shift how I worked. I knew I wouldn't be able to stop working altogether, but I wanted to do it differently—more tribal, together on the land. I have childcare now, but we live on the land where we work. They can call us on the walkie-talkies if they need anything. If someone falls and bumps their knee, we can be there to hold them. They can come down to the restaurant or the greenhouse, work with us for a little while, then go back to the comfort of the air-conditioned house.

This is a compromise. I don't believe you can be an exemplary mother and an exemplary entrepreneur—at least not fully, not at the same time. My life was already as an entrepreneur, and we're not set up for me not to work, so this is the best compromise I've found. I spent years being hours away from my children—an hour and fifteen minutes without traffic, longer with it. This feels better. Not to mention that my brother and other family members live on the property. Like a herd of animals, we look out for each other's children. They take their go-carts between each other's houses, play hard in the woods, and learn responsibility, and we're all here together. Even if I'm working or distracted, they always know they can come tug on my shirt and get my attention.

Nursing Nature

And then there's breastfeeding. The amount I've breastfed each of my children has depended on how much they needed, but all of them nursed at least until age two. My son Sol breastfed until he was three and a half. People are judgmental. It's interesting, the judgments I've gotten from people for freely feeding my children wherever I am, whatever I'm doing. There's the classic, "Do you want to do that in the bathroom?" No—why would I want to do that in the bathroom? Do you eat in the bathroom? I've had women, out of their own insecurity, walk up to me unprompted and say, "There's nothing wrong with formula." On social media, I've had people angry at me for posting while feeding my child—like it's some offense to their sensibilities.

I think we'd be better off if we normalized breastfeeding. The same amount of breast you can see while a baby is feeding—just the top, because the baby covers the nipple—we see all over advertisements and billboards. For some reason, it's okay to show the top of a breast as a sexual enticement, in a low-cut shirt with a push-up bra, but not when doing the most natural thing—the thing breasts were designed by God to do: feed children. I suspect if you feel weird seeing a woman breastfeeding her kids, it's likely not her problem—it's yours. We should normalize it, plain and simple.

But if you look at cows in a beef herd—because in dairy herds, humans pull the calves too soon—you'll see something interesting. The mothers allow their calves to nurse until the next one is born. Sometimes you'll see these fully grown calves, nearly the size of their mothers, still suckling. Beef farmers know that calves left on their mothers grow stronger, bigger, and faster than those pulled too soon.

I saw this truth come knocking—literally—on a frigid January morning a couple of years back. My neighbor Chuck rolled up in his Polaris, the engine rumbling against the quiet of the snow-dusted fields. In the back, bundled in a tarp, was a scrawny black Angus calf, rejected by his mother and trembling from the cold. Chuck climbed out, his boots crunching on the frozen ground, and nodded toward the calf. "Do you want him?" he asked. "Maybe you can get one of your dairy cows to take him." I didn't hesitate. "Sure," I said, and together we lifted him out. He was so small, so fragile, I couldn't leave him outside. I carried him straight into the living room, setting him down by the fireplace where the embers were still glowing from the night before. The Christmas tree stood in the corner—we still hadn't taken it down, and its lights cast a soft flicker over the scene. It felt right, somehow, to give him that warmth, that shelter. We named him Frío, a little tribute to the icy day he was born and abandoned.

Breastfeeding occurs anywhere—here in the greenhouse, with oranges for pest control among the lettuce.

We didn't mess around with milk replacer. From the start, we fed him colostrum—the first, rich milk from one of our cows—then kept him on straight milk every day, bottle after bottle. He'd suck it down, his eyes slowly brightening, though he stayed weak for a while. Eventually, we got one of our dairy cows to accept him—marginally. She was a patient one, already nursing her own calf, and she'd let Frío nudge in for a sip here and there—never enough to really fill him up, though. She'd tolerate him for a moment, then shift away, leaving us to pick up the slack with the bottles.

He survived, but thriving was a stretch. Months later, I asked Chuck if he wanted to see how Frío was doing. He just chuckled, shaking his head. "Don't need to look at him," he said. "He's short, pot-bellied, half the size of the rest of the calves that hit the ground that same day. I've seen it before." He was right. I'd already bought some of those other heifers from him—calves born around the same time, raised on their mothers' milk, grown out strong and tall, and sent off to processing. Frío, at a year and a half, is still squat and stunted—a living testament to what's lost when that natural bond breaks too early.

In commercial dairies, it's standard to pull calves right after birth, bottle-feeding them a ration of milk or milk replacer—or some combination of the two—while the mothers are hooked up to machines for our consumption. Much of this book is about what I've learned from nature or from farming, but in this case, I always knew I'd breastfeed, even before I got to the farm. I'd already nursed two children for extended periods by the time we started our dairy herd, so our approach here reflects that. We allow the calves to stay with their mothers. We don't get as much milk—nowhere near what a big operation pulls—but I believe if you're going to breed a cow every year and take her milk, the least we can do is let her raise her offspring, too. I'm not standing in judgment of other dairy farmers. Feed is expensive, margins are thin, and consumers want—or need—inexpensive dairy. I get it. I just wish there was a kinder way.

In a healthy beef herd, though, you'll see something different. Plenty of calves stay on their mothers until the day they head to auction, usually a month or two before she's ready to calve again. I've watched those big, lumbering calves—some nearly as tall as their mothers—still ducking under to suckle, their tails flicking contentedly. That milk, straight from the source, is the most nutrient-dense food on the planet for them. It's not just about growth; it's about health and resilience—everything they need to become what they're meant to be.

Yet here we are, humans, rushing to wean our own children like it's a race. In the United States, the average length a woman breastfeeds is three months—three months!—before we're shoving bottles and formula into the picture, convinced that's the milestone of progress. Do you know who one of the largest lobbies against paid family leave is in the United States? Formula companies. Do you know that we're the only first-world country with such a pathetic amount of paid family leave for mothers? Formula companies lobby against giving mothers more time with their children—for their profits.

Is there anything more diabolical than spending lobby dollars to separate a mother from the baby that still needs to be on her chest? We live in a world run by corporate greed. Our government shouldn't be swayed by corporations that sell seed oils, corn syrup, milk solids, and some Frankenstein version of what God has flowing out of every mother to make policy about family leave.

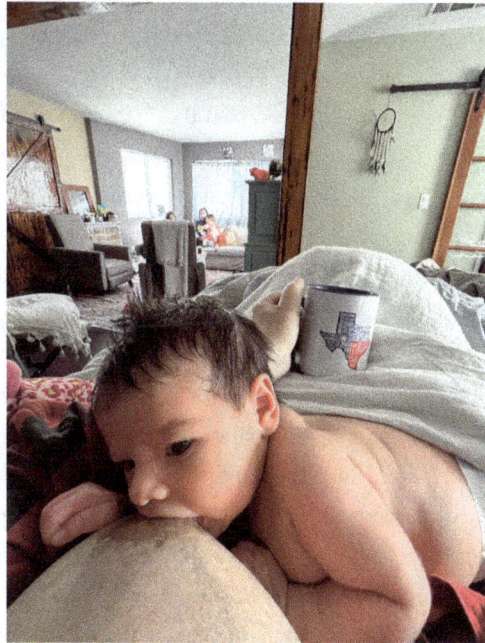

In bed after giving birth to Estrella, drinking tea—and dreaming of a future in Texas.

The numbers show that mothers start out strong—about 83 percent of infants born in 2019 began receiving some breast milk. At three months, though, only 60 percent are still getting any, and just 46 percent are exclusively breastfed. By six months, it's down to 55 percent for any breastfeeding, and 25 percent exclusive. A year in, only 36 percent are still nursing at all. Globally, many babies breastfeed for up to two years. What this tells me is that 83 percent of mothers want to breastfeed their children—they start out intending to—but something in our culture, our policies, our lack of support rips that away.

Why? What's the urgency? What are we running toward? No formula, no processed milk substitute, no expensive "next stage" baby food compares to the living, perfectly tailored nourishment of breastmilk. And beyond nutrition, breastfeeding is comfort, safety, security—another tether between mother and child that we've been told to sever too soon. It's obvious to every mammal on this planet except us: our offspring should stay on the breast as long as possible. It's obvious to them that babies should sleep next to us, too. But we've forgotten—or been taught to ignore—what our bodies know.

We have also been told that babies need to "self-soothe" and that leaving them to cry at night will teach them to sleep. But babies don't self-soothe—they give up. They stop crying because no one is coming. And what lesson does that teach them? Not resilience, not independence—but that their needs don't matter, that no one cares, that the world is indifferent. This is the first betrayal. This is where the deep, subconscious belief takes root: I am alone.

The idea of sleep training as we know it today didn't spring up overnight—it evolved, shaped by voices far removed from the rhythms of motherhood. One of the earliest influencers was Dr. Emmett Holt, a pediatrician from the early 1900s. In his book *The Care and Feeding of Children* (1894), he advised parents not to respond to babies crying at night, pushing rigid schedules over instinct. A man, a doctor, and likely not the one rocking a fussy child at 3 a.m., Holt laid the groundwork for what would come later. Fast forward to 1985, and Dr. Richard Ferber—a pediatrician and father—published *Solve Your Child's Sleep Problems*, popularizing the "Ferber Method." It's a form of graduated extinction: let your baby cry for progressively longer stretches before intervening. Both men, both experts, both shaping how we think about sleep from a perch that didn't involve the visceral, daily reality of caregiving.

It's no coincidence these ideas took root as the world industrialized. Back when families lived closer to the land, a mother's rhythm flowed with her child's—nursing on demand, sleeping side by side, adapting to the seasons. But as factories rose and cities swelled, women were pulled into a different grind. A rigid schedule became the backbone of survival: clock in, clock out, get the kids on a timetable so you can function at a job the next day. Sleep training fit into that machine—it promised efficiency, order, a baby who wouldn't derail the demands of a paycheck. It made sense for a world obsessed with productivity, but it clashed with the natural pulse of motherhood, the one that says stay close, listen, and respond.

These methods weren't born from biology or anthropology—they came from a culture prizing independence over connection, often dictated by men who didn't carry the weight of nurturing day in, day out. Today, you'll find sleep experts—many of them mothers or female pediatricians—pushing back, talking about attachment and the emotional echoes of letting a child cry. But the old ideas linger, baked into our parenting books and our guilt.

Prioritizing Presence

I never set out to be the mother with a double king bed and a decade of co-sleeping under my belt. It just felt right. And I didn't see the logic in disrupting something that was working simply because we were adding more children to the family. I'm grateful for our giant bed and the memories being created in it—snuggling our children, watching movies, tending to them when they're sick. In the blink of an eye, it will all be gone, and they will be grown.

Nursing Luna while reading her a story.

This does not mean I don't sometimes feel overwhelmed or overstimulated. There are moments when I need my body to be my own. As I write this, it's four in the morning. I'm downstairs, away from my children, drinking tea and getting these thoughts out of my head. I take moments for myself. But that's not the same as checking out of motherhood.

We've lost our way. We've forgotten the most natural, mammalian way of raising children—the way that prioritizes presence, touch, nourishment, and instinct. Modern parenting asks, "How do I get my child to be independent from me?" But nature asks, "How do I ensure my child thrives?"

The answer has always been the same. Stay close. Feed them from your own body. Answer their cries.

This isn't radical. This is what every other mammal on this planet still knows. We just have to remember.

CHAPTER 15

Lost Connection

I've felt that fracture in my bones. With Rio, my first, I fought for days—three brutal days of labor—resisting the hospital, the scalpel, all to give him that foundational health nature intended. The suffering was mine, not his: contractions racking me, my body shutting down, a broken coccyx, until they pulled him out with a vacuum and an episiotomy. Nurses swooped in to bathe him, to scrub away my microbes, but I stopped them. "No," I said, voice raw. "He stays unbathed."

For days, I kept him skin-to-skin, unwashed, clinging to what I could salvage of that primal handover. It starts at birth. When we slide through our mother's birth canal, her microbiota coats us—eyes, nose, mouth—kicking off our gut's ecosystem. It's a handoff, a gift of life.

But a C-section swaps that for the sterile sting of an operating room, and the baby's first microbial welcome comes from industrial ghosts, not Mama's living warmth. God's design, nature's design, has mothers give their babies what they need—microbes and all. On the farm, calves drop to the earth, and their mamas lick them clean, seeding them with the land's microbiology from breath one.

But us? We're obsessed with sterilizing everything—wiping down, washing off, bleaching out. We might be sterilizing ourselves into oblivion.

Half of Me Isn't "Me"

I've hashed this out in pieces for *The Epoch Times* and chewed it over on podcasts more times than I can count, but it's worth hammering home: we've lost connection. Not just to the land or our food, but to the raw, pulsing thread that ties us to life itself. Our gut microbiome mirrors the soil's living world, and we're torching both. Herbicides, pesticides, antibiotics—we're scorching the earth and our insides, killing the colonies that keep us alive. I'm Mollie Engelhart—mother, farmer, human—and I marvel at this body, this vessel for my soul. But half of me isn't "me"—it's the microbiota humming inside, keeping me going. Lose that, and I'm done—health, clarity, years, all of it.

Luna helping our goat Ginny nurse a kid she had rejected. Raw dairy has been a significant element in our effort to feed ourselves and our children nutrient-dense food.

Mothers who skip breastfeeding or lean on formula early might miss laying that same groundwork. That microbial base isn't just about digestion—it's our immune system, our mood, our skin, our whole lifespan. I nursed all four, through exhaustion and chaos, because I knew what was at stake—Rio's tender gut, Estrella's iron strength, all rooted in that connection.

After four kids, my body wouldn't budge—clean eating wasn't cutting it. I'd been diving into gut science, learning how kids have the richest stash of Bifidobacteria, thriving on breast milk. I was investigating this because I had a friend who felt like her gut had been ravaged by the COVID vaccine—no energy, no gut health, nothing left of her old fire. Studies trickled out: some

Luna was happy to return Ginny and her baby to the flock.

folks lost their gut diversity post-shot, Bifidobacteria wiped out. I wanted to help her, to find a way to rebuild what she'd lost.

That's when it hit me—raw milk might be the key. I'd test it on myself first. For eighteen days in Texas, I drank nearly a gallon a day—thick, creamy, straight from my Jersey herd. I'd sit on the porch at dawn, kids still asleep, the air heavy with quiet. No hunger, even with thousands of calories. Twenty pounds melted off, inflammation vanished, energy surged like I'd tapped some ancient well. My stomach flattened out, my cravings for carbohydrates disappeared, and even a strange little cyst on the bottom of my foot went away. The inflammation in my hands and feet that used to hit me every morning? Gone. I'm back in clothes I hadn't worn since before my kids were born. This might be TMI, but I still went poop twice a day—the first heavy, like my body was purging waste stuck for years; the second light brown, almost odorless, like a nursing baby's stool.

It was the most extraordinary fast I've ever done. As a breastfeeding mom, I hadn't fasted much in a decade—most aren't safe in that season. But this? No suffering. You might crave a bite of something else, but you're satisfied. The nourishment flows, the pounds drop, your energy climbs, and your body resets—restoring a healthy gut, just as nursing does for all mammals. I've done it since—once for the full stretch of Lent, 40 days, feeling unstoppable. I've been so excited I've started inspiring people—online and in person—to try it. The results are obvious and amazing. I hand jars to friends now—folks wrestling fatigue or bloat—and say, "Try this. It's life." On the farm, it's the same: calves suckling their mamas grow fierce and fast; bottle-fed ones lag. Connection's the key.

We douse our food with glyphosate, a weed-killer first patented as an antibiotic. Every bite of conventional bread or corn chips away at our gut's allies. From 1974 to 2014, the US dumped 3.5 billion pounds of it—19 percent of the world's total—on our fields. It's in our kids' snacks, our air, our blood. Some folks, desperate, turn to people living in traditional and indigenous ways—those who cook over fire, sleep on dirt floors, grow their own food—and buy their poop. They encapsulate it, swallow it, hoping to reseed what we've burned out.

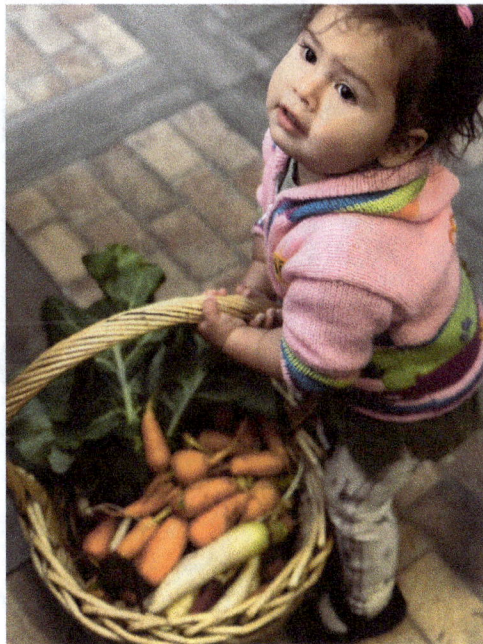

A younger Luna carrying fresh vegetables from the field to the house.

That's how far we've strayed. We've got to wake up. Fermented foods, raw milk, a wild mix of plants—some say thirty kinds a week—local pastured eggs, nutrient-dense meats, and bone broth feed those microbes. Bifidobacteria, Lactobacillus—they're the quiet heroes, breaking down fiber, fighting invaders, crafting vitamins we can't live without.

By Design

But it's deeper than milk or weight loss—it's how we're designed. Our microbiology, the intricate balance of life within us, reflects how we're meant to live, in harmony with God's creation. Breastmilk isn't just food; it's a divine blueprint, tuning an infant's gut to thrive. Raw milk, straight from the animals God placed in our care, echoes that intent—an untainted gift that realigns us when we've strayed. We're not separate from this living world; our bodies are woven into it, and through that, into God's purpose. The more

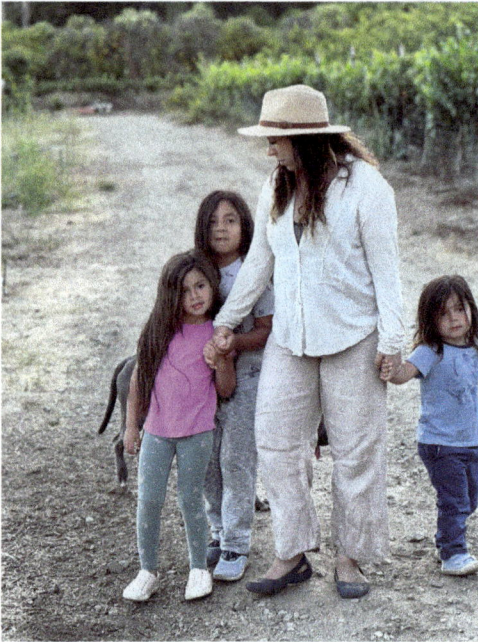

God whispers through every cell: "This is how I made you to be."

my body heals, the more I feel this truth: our connection to our microbiology is inseparable from our connection to Him. It's like He's whispering through every cell, "This is how I made you to be."

Losing that link isn't just about farmland or food. What if it's bigger? What if losing healthy soil—its microbiology, its pulse—cuts us off from creation itself? Soil isn't just dirt—it's the womb of life, birthing and breaking down everything. When we till it raw, drown it in chemicals, and monocrop it to death, we don't just lose nutrients. We're hollowing ourselves out—what I call the industrial microbiome—unplugged from nature, from the sacred flow that holds it all together. We're not just less healthy; we're less whole.

We say "trust your gut," "gut feeling," "go with your gut"—why? That second brain is more than nerves; it's creation singing inside us. What if shredding our soil rips us from God? What if those microbes are the divine breath, stitching us into coherence—body, mind, soul? I see the sickness swelling—diabetes, brain fog, kids lost to screens—and think: is this us forgetting we're part of the dance? On the farm, babies hit the ground and their mamas lick them into life, tying them to the land's heartbeat. But we scrub and sanitize, terrified of dirt. Maybe we're scrubbing God out, too. Without a gut alive with nature's chorus, can we hear those quiet nudges—the ones that tell us who we are, where we fit?

Regenerative farming is my way back—a remembering that we're not plagues but pieces of the whole. In Texas, we've turned cracked dirt green, pulled rain from dry skies, grown food so rich you feel it in your bones. But it's more—it's me woven into creation again. Every day, dirt under my nails, I know it: nature's true, unbroken, and I'm in it. Humanity's not the enemy; we've just lost our place. That lie—that we're apart, wrecking everything—leaves us lonely, unmoored. And what if that loneliness festers? What if it's why we're breaking—body and spirit—because we can't feel the earth's pulse anymore?

Science shows that gut chaos is tied to depression, infertility, decay—but I feel it deeper. My kids bloom here, synced to this land. The soil heals when we quit fighting it. Regenerative farming isn't just about crops or water—it's a lifeline back to God, back to belonging. We're not outside nature; we're of it. Believing otherwise has gutted us; it's left us barren inside and out. My prayer is that tending the earth mends us—reknits us to faith, to each other, to the whole we were made for.

A society that can't trust its gut can't trust itself—or the divine. That's when the government slinks in, choking us with rules—a government that bans my raw milk unless I bow to their game, that doubts two adults can sort their own deal. Only a people this lost would let that slide.

Our disconnection isn't just microbial—it's spiritual, and it's splitting us wide open.

CHAPTER 16

The Misguided Celebration of Excess: Body Positivity and Nature's Truth

In high school and college, I fought a silent, brutal war against my body, chasing an illusion of worthiness that only left me exhausted and empty. Obsessed with being thinner, I wrestled with eating disorders, trying to force my body into an impossible ideal.

It took years to unlearn that self-loathing and to embrace my body as a miraculous vessel—one that heals, creates, and connects me to the universe through microbiology, electricity, and even the emerging science suggesting water holds memory, per the work of Veda Austin and Gerald Pollack. We are, after all, mostly water. New research on our biofield hints at our power to shape reality through thoughts, speech, beliefs, and actions. Made in the image of the divine, our bodies are a living testament to creation's beauty.

Body positivity, in its purest form, should reflect this reverence: a call to love the sacred vehicle of our soul. But somewhere along the way, this noble idea twisted into something else. On platforms like TikTok, body positivity has morphed into a celebration of gluttony, a denial of metabolic health, and a glorification of conditions that harm us.

Calorie-rich, Nutrient-poor

I was recently a guest on the podcast of a remarkable young woman—Penelope Popken. At sixteen, Penelope weighed 320 pounds. By nineteen, she had lost significant weight, and she and her mother now teach other young women how to cook nutrient-dense, whole foods that transform their lives, through Penelope's company Step It UP. When I asked how she reached such a weight so young—especially with a health-conscious dietitian mother—she pointed to two culprits: TikTok and her school's food environment.

With Mimi at a wedding in Glendale, California. This was me at my skinniest as an adult.

"TikTok told me I was beautiful this way," Penelope said. Influencers and viral videos celebrated her size as empowering, drowning out any concern for her health. Her doctors, school counselors, and even her personal therapist affirmed her "perfection," offering no encouragement to address her wellbeing. Meanwhile, her school's lunch and snack options—laden with processed carbs, sugary drinks, and junk food—were available in unlimited quantities.

"My mom would never let me have that stuff at home," she explained. "But at school, I could get chips, soda, and pizza every day. It was everywhere, and no one stopped me."

Starving for nutrients while overdosing on calories, Penelope was trapped in a cycle of nutrient-poor food and societal affirmations that undermined her health. We failed her—not just as individuals, but as a society that confused affirmation with love. Of course, not every voice on social media denies the realities of metabolic health—but the loudest, most viral ones often do.

In nature, we see no obese wild animals. Deer, wild hogs, and wolves eating their native, whole foods don't balloon to unhealthy weights. Only when animals—like farm pigs or humans—gain unlimited access to processed grains or calorie-dense, nutrient-poor foods do they become metabolically unwell. This isn't about shaming; it's about observing what nature teaches.

A cow grazing on rich pasture sits contentedly by 10 a.m., chewing its cud, satisfied. On poor forage, it wanders all day, searching for sustenance. Our modern food system—mirrored in school cafeterias—offers the equivalent of that poor forage: shelf-stable, nutrient-deficient, leaving us hungry no matter how much we eat. For Penelope, school lunches and snacks were a daily temptation, undermining her health while TikTok and trusted adults told her she was fine.

All mammals prioritize future generations. In the wild, animals instinctively seek enough food, water, and shelter to ensure they can reproduce and sustain their offspring. A healthy reproductive environment is non-negotiable—obesity in wild animals is rare because it impairs fertility, mobility, and survival, threatening the species' continuation. Yet humans have strayed from this natural imperative. Obesity, driven by processed foods and sedentary lifestyles, creates a metabolically unhealthy environment that disrupts hormonal balance, reduces fertility, and complicates pregnancy. By celebrating a condition that undermines our ability to produce and nurture future generations, we are not just defying nature's blueprint—we are endangering our legacy.

The numbers tell a stark story. In 1980, the US adult obesity rate (BMI ≥ 30) was 13.5 percent. By 2017–2018, it had climbed to 42.4 percent, with severe obesity quadrupling in the same period. Among children and adolescents (ages 2–19), obesity prevalence doubled from the early 1990s to 2018, reaching 19.3 percent, with 6.1 percent classified as severely obese. Globally, obesity has nearly tripled since 1975. In 2022, over 1 billion people worldwide lived with obesity, including 159 million children.

These trends aren't just statistics—they are warnings flashing like wildfire. Obesity is linked to heart disease, type-2 diabetes, stroke, certain cancers, and reduced life expectancy. Yet instead of confronting this crisis, we're told to celebrate it.

The body positivity movement began with noble roots. In the 1960s, the fat acceptance movement emerged to combat weight-based discrimination. A pivotal moment came in 1967 with a "fat-in" in New York's Central Park, where activists protested societal bias by burning diet books and holding posters of thin models. By the 1980s, fat activism had spread globally, challenging the diet industry's scams and fighting for dignity. These early activists weren't promoting sickness—they were demanding respect.

The term "body positivity" gained new life in the early 2000s, amplified by the internet and social media. By 2012, it had exploded across Instagram and beyond, with hashtags like #bodypositivity and #loveyourbody celebrating diversity and challenging unrealistic beauty standards. It resonated with women, especially, who faced relentless pressure to conform. As someone who once starved herself to fit that mold, I understand the appeal.

With Elias in the kitchen at Sow a Heart Farm. I'm pregnant with Estralla here, and at my heaviest. Nature has been my greatest teacher in helping me understand what true health is.

Chasing Health, Not Thinness

But the movement has veered dangerously off course. Social media, particularly TikTok, has reduced body positivity to catchy slogans and unfiltered selfies, often normalizing obesity while dismissing health risks as "fatphobia." Tragically, this narrative has real consequences. In the last eighteen months, several prominent body-positivity influencers have died young, often from obesity-related complications—deaths largely met with silence from a media unwilling to confront the hard truth.

True love demands both acceptance and accountability; it asks us to honor the sacredness of the body, not betray it.

Nature has always been my greatest teacher, revealing truths about balance and health that no trend can obscure. When I announced my restaurants would shift from vegan to regenerative, embracing grassfed beef, raw dairy, local fruits, and leafy greens, I expected pushback. But I didn't expect the cruelty. Social media lit up with attacks on my appearance—an irony not lost on me. Acceptance, it seemed, was conditional on compliance with a certain ideology.

Until that moment, I had worn my extra weight as a badge of honor—the physical testament of four children born in rapid succession. Yet the media frenzy that followed—features on Fox News, Yahoo, The Daily Mail, and others—forced a deeper reflection. Thousands dismissed my insights on health and wellness because of my body, branding me "morbidly obese" despite my clean bloodwork and active lifestyle.

Still, I listened. I wanted my body to reflect the vitality I preached. Already eating clean and moving often, I turned to nature for answers. I embarked on the raw milk fast I mentioned earlier. Consuming over a gallon a day—roughly 3,000 calories—the weight melted away. The journey wasn't about chasing thinness; it was about honoring my body's design, restoring the vibrant health my children—and my mission—deserve. The "calories in, calories out" mantra we've been fed for decades misses the point. Nutrient density—not restriction—is the path to true health.

On the farm, obesity occurs for the same reason it does in humans: unlimited access to the wrong foods. Pigs fed endless grain grow unnaturally fat. Humans eating processed, shelf-stable foods suffer the same fate. Growing food reconnects us to its sacredness—the energy it takes to nourish life.

We are not separate from nature. Our bodies, like the deer's or the wolf's, are designed to thrive when given the nourishment they need. Celebrating sickness, even under the banner of self-love, will not build strong, resilient generations.

Our ancestors fought for survival—crossing oceans, enduring famine, surviving unimaginable hardship—to give us life. What will we hand to our children? Will we offer them a legacy of apathy and excess—or a legacy of reverence, resilience, and health?

Body positivity should mean loving our bodies enough to nourish them, strengthen them, move them, and respect their sacred design. It should mean telling the truth, even when it's hard. It should mean fighting for the future—not just affirming the present.

Let us reclaim the wisdom written into our very cells—the wisdom that knows health, strength, fertility, and love are inseparable. Let us build a culture that honors life itself.

CHAPTER 17

Equity, Equality, and Diversity—Nature's Truth vs. Human Ideals

E quity, equality, and diversity—these are words we hear constantly in modern culture. While much of this book is about how humans have strayed from their nature, I also believe that we are unique among animals. In that way, I suppose I am a speciesist. I value humans over other animals—and certainly over machines.

I've always found it fascinating that humans alone wrestle with the idea of valuing one species over another. In nature, a lion doesn't pause to consider the antelope's feelings before the hunt—it acts according to its instincts. Yet here we are, debating the ethics of eating meat or building machines that mimic our minds. This capacity for reflection, for questioning our place in the world, is what makes us extraordinary. It's not just compassion that sets us apart; it's the ability to imagine a world beyond survival, a world where we can choose to lift up the weak even when nature wouldn't.

What makes us extraordinary is our capacity for compassion. We have the ability to choose selfless acts—not just for our offspring but for others in our communities. That does not exist in nature—at least not in the same way. We can look at a situation and feel compelled to help, to contribute, to make things better.

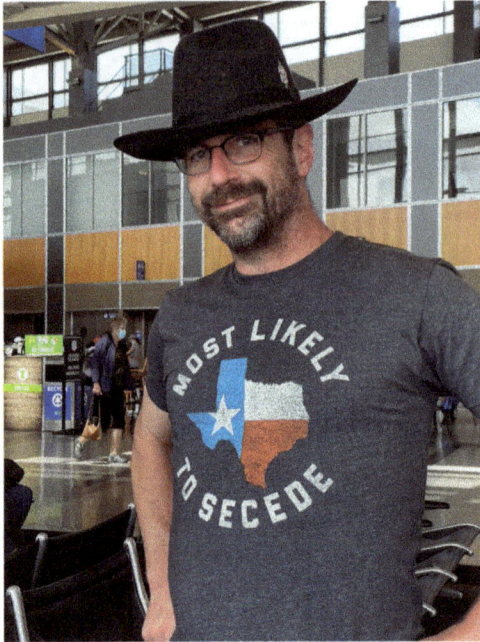

My long-time friend Trevor. He's my resident "no-man"—by contradicting pretty much every idea I ever had, he forces me, in a healthy way, to rethink my presuppositions.

But this uniqueness comes with a cost. In our pursuit of ideals like equality, we sometimes lose sight of the raw truths that govern life. I've seen this play out in my own journey—from the cramped quarters of my vegan ice cream shop to the sprawling fields of my farm. We want to believe everyone can do everything, but nature—and our own bodies—remind us there are limits. This isn't a call to abandon our ideals; it's a plea to ground them in reality.

Our moral compass and our understanding of a spiritual world beyond ourselves set us apart from the rest of the animal kingdom. The idea of equality is noble—something unique to human civilization. But it does not exist in nature. In the wild, only the strong survive.

I believe we should strive for equality, but we must also acknowledge that life itself operates on a fundamental truth: strength prevails. This doesn't mean we abandon the weak; of course, we care for those with disabilities, the young, the elderly, and the vulnerable. But we cannot pretend that all people can do all things equally, because that is simply not reality.

The Burden of Equity

I experienced this firsthand when I was opening one of my first brick-and-mortar businesses in Los Angeles—a small, 700-square-foot vegan ice cream shop. That little ice cream shop wasn't just a business—it was a dream I'd poured my heart into. I'd spent months perfecting vegan recipes, sourcing ingredients, and imagining a space where people could gather. But when the county inspector walked in and handed me a list of required upgrades, I felt the weight of idealism crashing into reality. Because I was investing more than a certain amount of money into upgrading the space, the county required me to make the entire store accessible for an employee in a wheelchair. The space behind the ice cream counter was a narrow strip—maybe three feet wide on a good day. I'd wedge myself in there, balancing scoops and spoons, occasionally banging my hip against the edge of the case. I'm barely five feet tall, and even I struggled to reach into the gelato case. When I was eight months pregnant, it became a comedy of errors: me, waddling and stretching, trying not to topple over while serving customers.

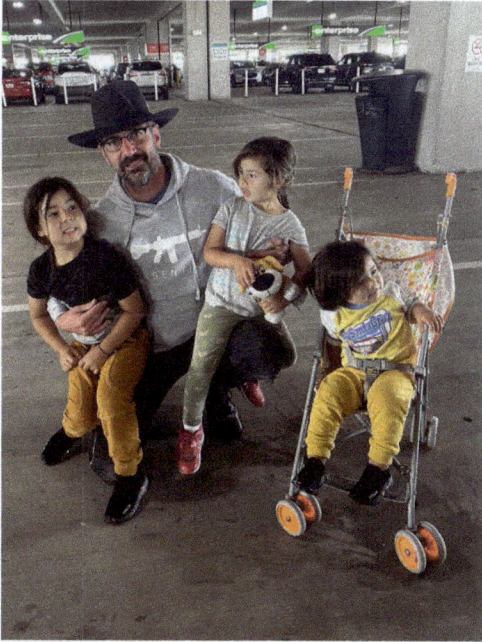

Trevor and my kids in the San Antonio airport. He came out to help us search for a Texas farm.

Now, imagine someone in a wheelchair navigating that space. It wasn't just tight—it was physically impossible without knocking over equipment or risking injury. I wasn't refusing to hire someone with a disability out of malice; I was looking at the tape measure and doing the math. Yet the law didn't care about practicality. It demanded I spend $40,000–$50,000—money I didn't have—on ramps, widened pathways, and modifications that would've eaten up half the shop's footprint.

I complied, of course, because I had no choice. But I couldn't help wondering: who was this helping? Was it equity if it forced me to compromise the business's survival? Was it equality if it ignored the job's physical demands? This wasn't discrimination. It was just basic logic. The same logic that prevents me from playing in the WNBA at five feet tall applies here: just because we want something to be equal doesn't mean it is.

I started asking bigger questions. How many small businesses like mine—run by people scraping by—get crushed under these mandates? How many jobs disappear because the cost of compliance outweighs the revenue? I'm not against accessibility; I want everyone to have opportunities. But there's a difference between opening doors and pretending every door fits every person. That experience taught me that equity, when divorced from reality, can become a burden rather than a blessing.

Of course, people should not be discriminated against based on race, gender, disability, or age. But we should also be honest about the reality that a 75 year old cannot do the same physical work as a 25 year old. On my farm, my uncle Scott—who worked hard as a mason his whole life—is now in his seventies. I recognize his wisdom, his experience, and his contribution, and I understand that he will lean heavily on the tractor or the Bobcat. When push comes to shove, he can haul hay or build fences. But at his age, he prefers to sit on the tractor, and of course we all understand that.

Yet, in the name of equity, we are constantly forced to pretend that everyone is equally capable of everything. This isn't just dishonest—it's counterproductive to a thriving society.

True Diversity

When we talk about equity and inclusion, we should remember that nature does not operate this way. In a flock of chickens, the weakest bird is attacked until it dies. In the wild, stronger animals eat first, while the weaker ones are pushed aside. However, what nature does value is diversity—but not the kind of superficial diversity we often discuss in modern culture. Walk into any thriving ecosystem—a coral reef, a prairie, a rainforest—and you'll see diversity in action. On my farm, I've watched this unfold in real time. The chickens cluck and strut, each with its role: the bold ones chase off predators, the quiet ones lay the best eggs. The goats browse the brush, keeping it from overtaking the pasture, while the dogs patrol the perimeter. Even the weeds play a part, holding soil together until something stronger takes root. None of them look alike, and none of them do the same job. Yet, together, they create a balance that no single species could achieve.

Contrast that with a cornfield—a monoculture stretching to the horizon. It's efficient, sure, but it's fragile. One pest, one drought, and the whole system collapses. Nature doesn't bet on sameness; it thrives on difference. A hawk doesn't try to swim like a fish, and a fish doesn't soar like a hawk. They complement each other—not by being equal, but by being distinct.

People tend to think of diversity in terms of race and sexual orientation. But in nature, diversity isn't about appearances—it's about function. Take a forest, for example. If you walk through a redwood forest in Northern California or the dense juniper thickets in Texas, it might appear as though there's only one species dominating the ecosystem. But if you look

Nicholas Skinner, another long-time friend who's always willing to (helpfully!) disagree with me. This was at the opening of our Pasadena restaurant.

closer, that's never the case. Nature—and God—never favor a monoculture. Instead, a healthy ecosystem is one where every plant and every animal plays a role, contributing something unique to the whole.

In human society, we've flipped this on its head. We've taken diversity—a principle of strength—and reduced it to a checklist. I've sat in meetings where the focus was on hitting quotas: X percent women, Y percent minorities, Z percent whatever else. But what about the diversity that keeps a team alive? The mechanic who sees problems I miss, the accountant who crunches numbers I'd fumble, the dreamer who envisions what I can't? That's the diversity I've learned to value—not the kind that's skin-level, but the kind that's bone-deep, rooted in what we bring to the table.

The previous US administration believed it was the most diverse in history—but in truth, it was one of the most monolithic administrations we've ever had. That's because nearly everyone thought the same way. What good is an administration filled with people who look different and have different sexual preferences and proclivities, but all think exactly the same? That is not diversity.

Real diversity means bringing together people with different experiences, different ideas, different perspectives. It means being willing to engage in conversation with people who disagree with you—not silencing them. It means ensuring that we aren't living in an echo chamber, surrounding ourselves with people who validate our existing beliefs rather than challenging them.

In business, I always keep a close friend or two who know they can't be fired for telling me the truth. Nicholas and Trevor have played this role for over a decade. They may not like telling me the truth. It may be awkward or uncomfortable to do it. But they do it without fear of retribution. I don't always agree with them. Often, I disagree with them. But having them push up against my ideas, against the way I want to move forward, always makes me stronger. It makes me a better entrepreneur and a better businessperson. Because I'm not surrounded by yes-men who agree with me, I can avoid pitfalls I might not otherwise see.

Trevor consistently found places people were stealing from us. His mind works in a way that mine does not. I would not be investigating the way he would. Nicholas pushes back against my ability to have it done and requires me to look at the long term and make decisions for quality when it comes to the aesthetics of our businesses. These are just two examples.

We should also engage with people who are politically different from us every once in a while. I have a dear friend, Rikarlo, and his wife, Jessica. Rikarlo and I met when we were barely out of our teens, two stubborn souls thrown together by chance. Back then, I was the bleeding-heart liberal and he was the guy who'd argue just to see me squirm. I'd rant about systemic injustice, and he'd counter with some half-baked theory about personal responsibility—usually while sprawled on my couch, eating my food. One night, with his foot in a cast and nowhere else to go, he launched into a tirade about how the white man was holding everyone back. I shot back, "Well, this white woman's driving you to Walgreens tomorrow, so maybe lighten up." We laughed, but it stuck with me: even in disagreement, we cared enough to keep talking.

Over the years, my views shifted—slowly, then all at once. I went from marching for causes to questioning them, from voting blue to voting red. Rikarlo didn't follow me there. He's still in his NPR bubble and still thinks I'm half-crazy for supporting Trump. But he didn't ditch me. Just last week

Me and my friend Rikarlo. We've disagreed with each other—and loved each other—for over twenty years.

we were at it again—two hours on the phone, debating everything from race to economics. He threw out that line about me being "blacker than Candace Owens," and I nearly choked on my coffee. It was absurd, but it was us: poking holes in each other's ideas, not each other's worth. Jessica, his wife, gets it too. She'll roll her eyes at our bickering, but she's the glue—reminding us that friendship isn't about agreement.

What I've learned from them is that real diversity isn't just tolerating difference; it's wrestling with it. It's sitting across from someone who sees the world upside-down from you and choosing to listen anyway. In a culture that's quick to cancel, that's radical. It's taught me patience, sharpened my arguments, and kept me humble. Rikarlo is one of many friends that I've had for over 20 years. And that's the real test—can you stay close with people as your life changes, as your views change, and as you see the world differently?

The importance of loving people we don't agree with is immeasurable. We must be able to separate the being—the expression of God in front of us—from the ideas that being holds true. We are not our ideas. I'm the same person I always was, and yet this entire book is about how my ideas shifted drastically over the years.

Shifting Goalposts

Unfortunately, much of the equity conversation is pushing us further apart rather than closer together. The ever-moving goalposts should always be a warning. When I was younger, equality meant fair treatment—same rules, same chances. Then it became equity: different rules to level the playing field. Now it's something else entirely—reparations, dismantling systems, rewriting history. Every time we think we're close, the goalposts move again. It's like chasing a mirage in the desert: the closer you get, the farther it slips away.

On my farm, I see what happens when you overcorrect. If I give extra feed to a weaker goat, it might thrive—but if I keep adjusting, soon the strong goats starve, and the whole herd suffers. Balance matters. Yet in society, we're told to keep pushing, keep tweaking, until everyone's outcomes match. I've watched friends in corporate jobs twist themselves into knots over this—hiring not for skill but for optics, promoting not for merit but for narrative. It's exhausting, and it's not working.

The data backs this up. In education, we've spent decades pouring money into closing achievement gaps, yet they persist—sometimes wider than ever. In business, diversity mandates often lead to resentment, not unity. I'm not saying we should stop caring about fairness; I'm saying we need to ask: is this helping? Or are we just pretending it is because it feels good? Nature doesn't reward endless adjustment—it rewards adaptation. Maybe it's time we adapted too.

True diversity isn't about quotas. It isn't about checking a box. It isn't about race, gender, or identity politics. True diversity is about bringing together thoughtful individuals with different strengths, different skills, and different perspectives—who all contribute to the whole. That's how nature thrives. That's how civilizations thrive. And that's the kind of diversity we should actually be striving for.

Equity, equality, diversity—these words carry weight because they reflect our highest hopes. But if we're honest, they've also become traps, luring us into a fantasy where strength doesn't matter and differences don't exist. Nature teaches us otherwise. It shows us a world where every creature has a role—not an equal one, but a vital one. On my farm, in my businesses, in my friendships, I've seen this truth play out: we thrive when we embrace what each of us brings, not when we force sameness.

So where do we go from here? I think it starts with courage—courage to say what's real, to value people for their contributions, not their categories. It means building teams, communities, and even nations where disagreement isn't a threat but a strength. It means loving the Rikarlos in our lives—not despite our differences, but because of them. If we can do that, we'll get closer to a society that doesn't just survive, but flourishes—like a forest, not a factory.

That's the diversity I'm chasing. That's the world I want to leave behind.

CHAPTER 18

Defund the Cartels

I f you've made it this far, you've noticed that a lot of the things I used to believe have shifted. Many of them radically. I'm not here to apologize for the change—I'm here to explain it. This chapter may be the most unexpected one yet, because nothing in my life would exist without illegal immigration. And yet, I now advocate for border security. Or, more explicitly, I advocate for a balanced view on this issue that would maintain the rule of law but that would recognize the reality of our labor situation and would allow America to thrive.

People ask me all the time, "How can you be conservative? Your husband came here undocumented. Your whole life is because he broke the law." And yes, that's true. I am deeply, unendingly grateful for that 16-year-old boy who borrowed money from his neighbors and took a treacherous journey north— not for selfish ambition, but to support his family. That brave decision gave me everything I cherish: my marriage, my children, my community. It gave me my life.

But while his courage is heroic, that can't be the model we use to shape national policy. A society cannot run on exceptions. We cannot continue importing millions of people from around the world and pretend it has no

When I started studying with my husband for the citizenship test, I gained a new layer of appreciation for America and for patriotism.

cost—on culture, on infrastructure, on cohesion. We act like our abundance is infinite, like there will always be more room, more money, and more opportunity. But much of America's abundance exists because we are the world's reserve currency. We print money out of thin air and buy goods with it. That illusion of unlimited wealth doesn't translate to reality when it comes to land, housing, or jobs.

On Borders and Fences

At first, I wasn't even sure I wanted to include a chapter about immigration in this book. After all, this is a book about how Nature is a conservative—about how the natural world reveals truths that our modern ideologies often try to ignore. And when you first look at nature, immigration doesn't seem to fit. You see migrating butterflies, migrating birds, wildebeests crossing continents, animals moving freely across the world—unbothered by borders. It's easy to think, "Well, maybe territory is a manmade idea. Maybe fences are unnatural."

But the deeper I looked, the more I realized the opposite is true.

Yes, many animals migrate. But migration and territory are not oppo-

Osmar, Mollie, and Osmar's attorney outside of the courthouse after we were granted guardianship.

sites—they're part of the same truth. Lions have territory. Chimpanzees have territory. Wolves and cougars and coyotes and even domestic cats have territory. They patrol their borders, they scent-mark trees, they howl or growl or fight to protect what is theirs. Chimpanzees have even been observed going to war over the most abundant feeding grounds. Male lions will fight to the death to protect their pride's territory from invading males. Dogs will bark, snarl, and bite when another pack tries to move in. Even a sweet barn cat will claw the eyes out of a rival who dares enter its zone.

So, while on the surface it might seem that immigration doesn't belong in a book about natural law, I now believe it absolutely does. Because territory is natural. So is protection of scarce resources. So is prioritizing your own family, your own community, your own offspring. No species survives by endlessly giving away its food or watering hole to outsiders. That doesn't mean those outsiders are bad. It just means nature demands discernment. It demands structure. It demands truth.

And what's true in the animal kingdom is also true for us. I often find myself daydreaming about immigration solutions, fantasizing that the President will call me up and ask, "What should we do about the border?" I'm not kidding—I actually have a ten-point plan typed up, ready to go, because I think about this issue constantly. Not just as a farmer and a chef or as someone married to a formerly undocumented man, but as a mother, a citizen, and a woman who sees both the immense beauty and the devastating consequences of the current system.

I've come to understand this: Nature allows for movement within the limits of structure. The birds don't migrate in

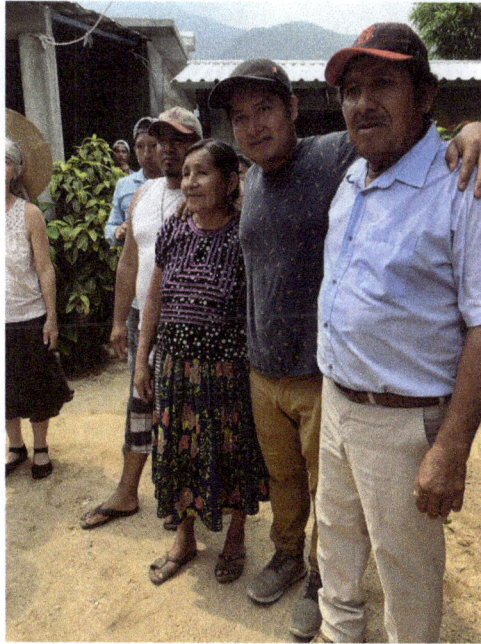

Elias was able to return to Mexico to finally see his parents again after more than eleven years.

chaos. They follow patterns—routes that are shaped by weather, food, season, and survival. And when a food source is threatened, animals defend it. When a species is vulnerable, it retreats inward to protect itself. This is not cruelty. It's wisdom. It's instinct. It's order.

So no, a secure border isn't unnatural. In fact, it may be one of the most natural things a nation can do.

Secured by the Cartels

There's another cost no one talks about: the dilution of what makes America special. Not skin color. Not religion. But the fire in our gut—the belief in freedom, in individual responsibility, in speaking out even when it's unpopular. Ironically, even the radical left—whose beliefs I've moved away from—operate with the assumption that we have the right to rage against the machine (although it seems nowadays the left is raging with the machine). That instinct is American. But not every culture holds it.

Cultures that come from authoritarian regimes or collectivist mindsets often bring with them deeply ingrained habits of silence, compliance, or tribalism. That doesn't make them bad people. But when assimilation is no longer encouraged, and when American values are seen as oppressive instead of aspirational, we risk losing the very thing that immigrants used to come here to embrace.

I'm not on one side of this fence. I live it. My husband, my community, and Osmar—who I brought into our family—are all deeply tied to the immigration story. When I was granted custody of Osmar, we walked out of that courtroom and I turned to him in the car and said, "We only raise patriots in this family." He smiled and said, "I know." What made that moment kind of funny is that he was already nearly a full-grown man at the time. But still—he knew what I meant. That in our family, love for this country isn't optional. It's a core value.

Here's the hard truth: the southern border is not secured by the United States government. It's secured by the cartels. Every person crossing is paying $10,000 to $13,000—often more—to make the trip. They are pre-scheduled to be robbed en route, and often extorted again once they arrive. The cartels run a billion-dollar operation, and we fund it every time we eat inexpensive produce picked by someone who had to cross illegally to get that job.

And we do need that labor. Desperately. Anyone who tells you otherwise hasn't run a restaurant, harvested a field, or worked in a meat packing plant. We've raised a generation of weak-bodied, weak-minded young people who believe hard work is beneath them and discomfort is a trauma. Our economy is addicted to the labor of immigrants—but our policies pretend we're not. And that dishonesty is killing people.

Many talking heads in the conservative movement rail against immigration, while much of their lives are supported by cheap labor. Even conservative farmers use that labor—quietly. What nature has taught me again and again is that we must be grounded in truth and in reality. And the reality is that our dependence on foreign labor—combined with our unwillingness to create a real policy and a sensible pathway—is leading to profound suffering.

Here's one of the most broken parts of our current system: there are almost no viable legal pathways for low-skill laborers from Mexico to come here. Some people from Central and South America can claim asylum and remain in the US while they await their hearing—a process that can take five to ten years. Even if their claim is ultimately denied, most never leave. But Mexican nationals—our direct neighbors to the south, also part of North America—

do not have that same option. We've created a system that makes it easier for people from distant countries to come and stay than for our closest neighbors. That makes no sense. We should be prioritizing Mexico first, then Guatemala, El Salvador, and Honduras. A strong neighbor is good policy. A secure, thriving Mexico is national security.

We see some of the suffering in headlines—dead children in the Rio Grande, families in the

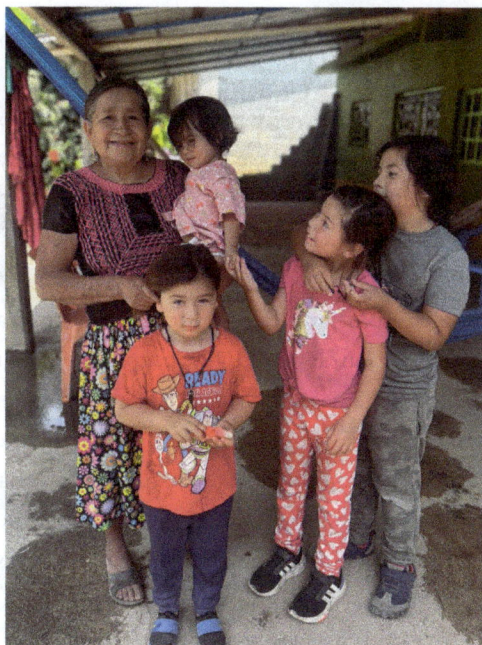

My children visiting their grandmother in the summer of 2024.

back of semi-trucks. We hear about women taking Plan B in preparation for rape. But we rarely acknowledge the quiet, invisible suffering: the man who's paid taxes for twenty years and hasn't seen his mother at all during that time. The woman who's missed every birthday of her children back home. Or the field worker who goes home to die of cancer, a gift from the chemicals we allowed him to touch but never tracked. These are real people. I know them.

We can do better. Humanity can do better. Christians can do better. Americans can do better. But right now, both sides are more committed to a position than to a solution. Everyone is more loyal to ideas than to progress.

A Practical Solution

Here's the truth we have to face: we need the labor. A border is not racist. A wall is not racist. I have a fence around my property—not because I hate my neighbors' cows, but because I don't want them eating my grass. And yet, this is a constant issue. My neighbor's cows get through my fence. They eat my grass. I fix the fence. It's not because I dislike their cows. Their cows are beautiful, majestic beings, just like mine. But I don't want to feed them—unless, ultimately, they are going to contribute to my family's life at the end of their own.

In the same way, we want people in this country who want to contribute to this country. And we must create a pathway for those who want to come, work, and then go home to their families.

That's why I've proposed a ten-year low-skill work visa. It would allow people to work, pay taxes, visit home, and live in dignity. It would protect women from rape, children from trafficking, and all migrants from murder and extortion. It would give people a legal alternative to risking their lives and their bodies. Applicants would need a clean criminal record. They'd be required to stay employed. The visa could be renewed once, but would never lead to citizenship—not even through marriage. The best case might be a green card. The goal is contribution, not permanent relocation. There would be a fee of $10,000 per person—meaning that same amount of money people currently pay the cartels would go to the US government instead. The grip of those cartels on society would be weakened.

Unlike our current system, it would allow people to keep connection with their home—so they're not forced into a forever exile from the people who raised them. It would dismantle the business model of the cartels overnight. It would allow people to fly directly into the cities where work is needed—no more desert crossings, no more smugglers, no more horror stories. Just honest labor for honest pay.

We must also learn to listen to people we don't agree with. I'm responsible for failing in this myself. Earlier in this book, I wrote about how my perception of Tucker Carlson was shaped entirely by what I thought he believed about immigration. But when I started really listening, he helped me understand why a secure border is important. And if I ever have the opportunity to sit down with him, I hope I can help him understand why our labor needs are just as important. Maybe together we could have a conversation that moves us toward real solutions rather than just amplifying rhetoric. No one will be 100 percent happy with any solution that gets put forward—but that can't stop us from trying. We the people must demand better.

We must also do better—and learn to listen. Tucker Carlson does not hate my children. He doesn't think they're replacing white children. He's an American patriot who cares deeply for this country and wants accountability. He's not afraid of being called racist for speaking his truth, and that's admirable. His truth and my truth overlap. With over fifteen years in hospitality and farming, I of course have many more immigrants in my day-to-day life, and their stories have shaped my heart and my convictions. But the more I listened to Tucker, the more I understood that he, too, is a compassionate and loving American. We must come together—even with those we don't fully agree with—to create something better.

The system we currently have isn't broken. It's working exactly as designed. The powers that be—the ones we cannot see—are profiting from the chaos. And for some reason, our elected officials have shown no real interest in designing a system that actually supports human beings. We see this same pattern in so many other areas of life.

But here's the truth: we are the ones we've been waiting for. You—reading this book. Mothers doing extra research to take care of their children. Farmers questioning government protocols. Citizens who recognize that something is deeply wrong, that we have gone astray. We must come together for solutions. Yes, we all know there's such a thing as "planned opposition." But we've gotten so good at judging and separating and canceling each other's ideas that we don't even need outside opposition anymore—we divide and conquer ourselves.

So here I am, laying out ideas that I know are imperfect. I hope they spark new ones. I hope they inspire thought, dialogue, debate, and even disagreement. Because that's how we build something real. And maybe, just maybe, together we can come to a more perfect path forward.

CHAPTER 19

The Whispers of a New ~~Horizon~~

I loved my farm in California with a fierceness that's hard to put into words. Sow a Heart Farm, as we called it, was my sanctuary—a slice of Eden nestled in Fillmore, where Sespe Creek sparkled through the property, misty mornings greeted us, and the soil we'd nurtured poured forth abundance. Most days hovered around 75 degrees, perfect and golden. I bought two nearby properties for my three kids, dreaming they'd inherit a legacy tied to this earth, all of us rooted close together. I thought I'd die in that house, surrounded by the community I adored, the land I felt stitched into my soul.

But whispers—soft and insistent—began to stir, hinting at a different path.

The first came from Blanchard, a friend as colorful as he was wise. A black man from Sacramento, now living in Atlanta, he is a patriot, a Christian, and a conspiracy theorist with a knack for seeing what others missed. One morning, as we sat at my kitchen counter sipping fresh-squeezed orange juice from our citrus trees, he fixed me with a look.

"Mollie," he said, "this thing you've got going here? It's amazing. But it's in California."

"Yeah," I laughed, "where else would it be?"

He chuckled, then grew serious. "The rural redneck is the last great hope for America. When the shit hits the fan, you'll look left, you'll look right—and there won't be anyone to fight alongside you."

I raised an eyebrow, amused but curious. "What are you getting at?"

"Texas," he said simply. "You, me, others like us—we should move there."

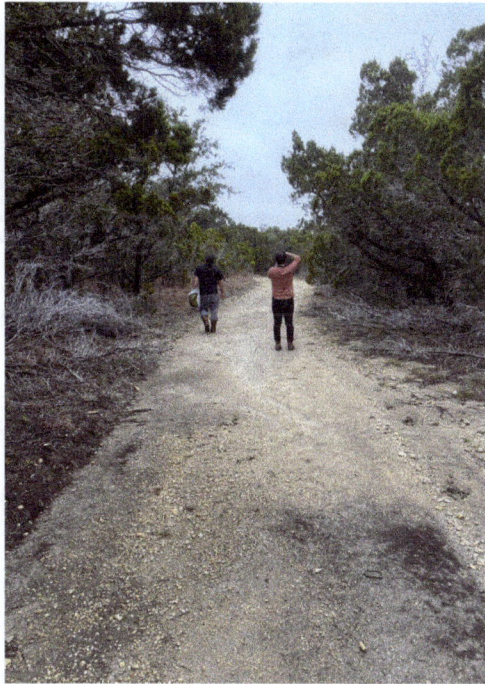

Nicholas and Elias scouting a farm in Texas.

I brushed it off as Blanchard being Blanchard. We toured the farm, hugged, and he left for Sacramento. But the whisper lingered.

Days later, at my Culver City restaurant, a regular approached me. "I've got a "Men's Warehouse" in Austin," he said. "You should put a Sage there." Two mentions of Texas in seventy-two hours. Odd, but not enough to sway me.

Then my company's COO, Steven Cook, sent me an article: restaurants were floundering in states like California under tight regulations and high wages, while Texas was a haven where they thrived. A third whisper.

A month later, on a family vacation in San Diego, the Airbnb owner—an entrepreneur with Mexican restaurants in Dallas—chimed in. "It's like printing money," he told us while skimming the pool, describing his bustling table service and late-night music scene. Four whispers, all pointing east. I'd always been good at listening when God spoke, so I booked a trip to Texas—not to move, but to explore. Maybe I'd open restaurants, expand my footprint, and sell later for a profit. California was still home.

In Texas, we scouted properties and prospective restaurant locations, then

Our new home in Texas—the day we closed escrow.

started looking for farms that already had the infrastructure for a restaurant. First, near Austin, we found one with an event space too, but it came with baggage—previous owners in foreclosure living next door, their resentment tainting the vibe. The second, an olive orchard outside San Antonio, boasted infrastructure but loomed under a cell tower I couldn't stomach raising my kids beneath. I measured its distance from the house on Google Earth obsessively until a cash buyer swooped in, freeing me from escrow. A blessing in disguise.

Then came Utopia—a remote paradise with flowing water so exquisite it stole my breath. We put an offer in, captivated. On the drive back, with baby Sol screaming in his car seat, I pulled into a Lowe's parking lot in Bandera to nurse him. Glancing up, I saw a sign for a property I'd eyed online—overpriced, impractical, but suddenly close. My dad, who'd joined the hunt, shrugged. "What we just saw in Utopia's pretty remote already." Curiosity won. We visited the Bandera place the next day.

It was a mess: 5,000 square feet, two bedrooms, one closet, 20-foot ceilings, and a single kitchen cupboard. The German builders had lost steam—trim dangled unfinished, plumbing reeked of sewer. I wasn't impressed. We piled back into the car, set on Utopia's magic water.

But at 4 a.m., a voice jolted me awake: Go back to Bandera. Grab a shovel. Check the soil. Sleep wouldn't return. Downstairs, I found Dad brewing tea in the Airbnb kitchen.

"What are you doing up?" I asked.

"I just woke up," he said. "We need to go back to Bandera and dig in the field."

The next day, we returned. The house didn't matter—what caught me was 70 acres of rock-free fields, a rarity in the rugged Texas Hill Country, with six to ten feet of rich soil begging for life. A creek carved through it, its banks revealing the earth's depth. The property stretched to 256 acres—or so the listing claimed—but it was over budget. Dad urged me to make an offer anyway. "Figure it out later."

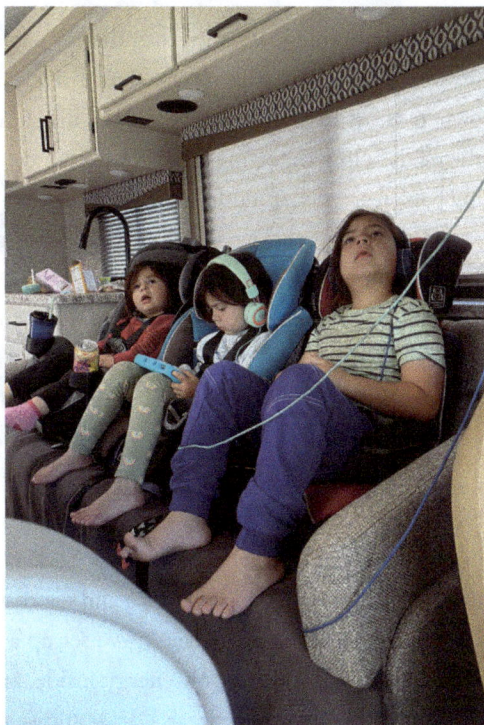

We packed the whole family into the RV to go close on the property.

We did. They accepted. Back in California, I pored over Google Earth, plotting my next farm. But the acreage didn't add up—my measurements topped out at 210. I requested a new plat map, negotiating that if it varied significantly from 256, we'd adjust the price per acre. It came back at 206, dropping the cost into reach. I laughed to myself: God had saved this place for me, hiding it behind a half-finished house no one else could envision. Even the unpaved 3.5-mile road to the property shrank to half a mile on our route—and during escrow, the county paved it entirely. Another gift.

One downside hung over us: an easement through a neighboring rancher's land. My husband bristled at it. Though the neighbor couldn't block us, we wanted his blessing. So we drove out, bearing beer and goods from Sow a Heart Farm to show him who we were. I flew in with my brother's family, my family, and my nanny's family—our full crew, part-time or not—to meet Chuck. He welcomed us warmly, intrigued by the brewery, the beer, and the farm's bounty. His excitement bolstered our resolve to close escrow.

For the signing, we piled into the RV—me, my husband, our three kids, my dear friend Nicholas, my assistant Brenda, and nanny Lupe. The previous owners let us park in the driveway the night before closing. As we wandered the fields below, dreaming of what we'd build, my phone rang. It was Bronson, my banker from Capital Farm Credit.

"There's a hitch," he said. "We're missing papers. Self-employment stuff. I'm not sure we can close tomorrow."

I hung up, looked to the sky, and whispered, "God, this is up to You. If it's meant to be, let it close. If not, we'll head back to California." Within a minute, Bronson called back. "Miscommunication. We're good to go."

Relieved, we crammed into the RV—three kids, four adults—ordered pizza, and watched a movie. Morning brought a surprise: snow. Just an inch, but in Texas, it shut everything down. The escrow office closed. Stores, laundromats—nothing opened. As a New Yorker, I was baffled. Stranded in the driveway, we waited out days of ice and wind. The previous owners eventually let us pull into a barn for shelter. We set up a folding table with an espresso machine, let the kids run off energy inside, and stretched our supplies with Walmart socks and underwear. DVDs and snacks filled the blurry hours.

Finally, escrow reopened. We signed the papers, and that night we stepped into the house. It was empty, reeking of sewage, so we slept in the RV again. I raced to Home Depot before closing, grabbing 4-inch caps to seal two open drains the previous owners had stuffed with plastic bags and Glade air fresheners. Standing there, caps in hand, I wondered: Is this really what I'm meant to be doing? Fear gripped me.

But we pressed on. My LA house-flipping crew flew out to tackle the Bandera house. We sliced it in half, added a 2,500-square-foot second story for my family, turned the odd original bedrooms into guest quarters, and expanded the kitchen into a proper heart of the home. Over months, we shuttled between California and Texas—planting trees, erecting a greenhouse, hiring a couple, John and Shay, to live on-site with my uncle Scott in his RV. I still saw it as a foothold, not a future. I'd open restaurants in Austin and San Antonio, supported by this farm, and keep my California life intact.

Then my friend and former employee Nicole pointed out something extraordinary: our Texas farm sat exactly in the path of totality for two solar eclipses, six months apart. It felt like God had drawn an X in the sky, whispering "This is where you're meant to be." As an entrepreneur, I couldn't let the miraculous pass by. I called Alec Zeck, a friend I'd collaborated with on past events. "Want to do a festival for each eclipse?" I asked. "Gatherings to bring scientists, musicians, doctors— like-minded souls—to this spot where the heavens say X marks the spot?" He brought his wife and kids to see the land, and his excitement matched mine. I called our mutual friends, Mike and Lindsey, and the Confluence Festivals were born. We dove into planning with fervor, envisioning a way forward for humanity on this sacred ground.

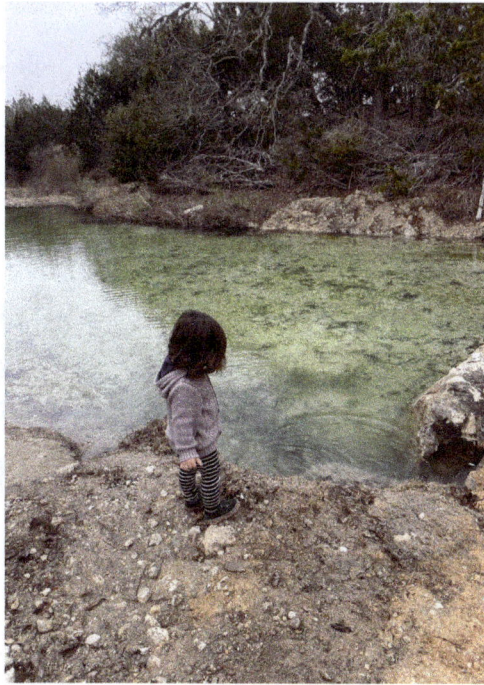

Sol throwing rocks in the crystal-clear creek at our new home. Our future begins here.

But as we prepared, my California businesses crumbled. The restaurants in Culver City and Agoura Hills closed within months. Like many restaurants around that time, regulations, costs, and COVID did them in. As pandemic restrictions began to be relaxed, customers expected full staff and full service, but with such low sales this was just impossible. We cut everything we could, and when we couldn't pay bills, the first thing to go was my and my husband's salaries.

We also faced difficulties from the state concerning our farm. The land had been farmed since 1884, and much of the infrastructure was old. The state's intervention began when we allowed a homeless family living in an RV to stay on our land. It's illegal in Ventura County, California, to live in an RV anywhere other than an RV park. A neighbor told on us, and the county came in. They walked the entire property and flagged every single thing that was not up to modern code — which was most of it. They wanted to evict all of our farm workers and told us we had to redo all the electrical and plumbing. We were using buildings for agricultural use that they said were only for automobile storage. The list went on and on. They slapped us with half a million dollars in violations and wanted to charge us $1,000 a month until we fixed everything.

My politics were shifting too, aligning more with Elias's conservative roots. Texas began to feel like more than a venture—it felt like a calling. We listed Sow a Heart Farm. I prayed daily for an all-cash buyer to take it off my hands, violations and all.

My father helped us build a greenhouse on our new property in Texas. It's such as blessing to have his encouragement and support!

One day, on my Texas porch, my phone rang.

"Hey, it's Ye," a familiar voice said.

"Ye?" I blinked. Kanye West, whom I'd met years ago, was calling about my farm.

"It's all about food, water, shelter," he said, his passion crackling through the line. "Shelter includes clothing too."

He toured the property, inspired by its regenerative spirit. "I'll write you a check right now," he offered. But his team stalled, and he soon jetted off to Japan for an album. The deal fizzled. I kept showing the farm, raising money for Texas—a brewery, a restaurant, tiny houses—until, during the first Confluence Festival in October 2023, an all-cash offer arrived. The buyer, heir to a pharmaceutical fortune, promised to keep my workers and the farm's mission alive. I laughed at the irony—God had answered every prayer, just not the way I'd pictured.

Escrow brought a new hurdle: moving three-quarters of a million dollars in equipment—greenhouses, tiny houses, animals. It cost thousands per trip. With weeks to go, bank accounts empty, I knelt in prayer. "God, if you could drop a pile of money on me, I'd really appreciate it."

At 3 a.m., my phone pinged. Ye, responding to a month-old text about Texas, wrote, "No, I'm not investing in anything that's not me." I thanked him, deflated. Then: "But I can hire you to consult." He pitched a farm project in Italy and told me to invoice him $300,000—the amount I'd once sought for Texas. Skeptical, I sent it. Days later, after a follow-up, the wire hit. That miracle funded our exodus—Osmar, Trevor, my husband, my brother Matthew, and countless others hauled our life to Texas. By December 19, Rios's ninth birthday, we were settled, starting anew.

The whispers had led me here—through loss, faith, and unexpected grace—to a horizon I never planned, but always needed.

CHAPTER 20

How Do We Reconnect?

Reconnection does not happen all at once. It's not a grand event or a singular moment of awakening where the sky splits open and everything becomes clear. It's not a Hollywood montage with swelling music and a tidy resolution. It's a slow return, a remembering, an unfolding—like the way a seed cracks open in the dark, reaching for light it can't yet see.

In that return, we must be careful not to carry the misguided burden of "saving the planet." That phrase has been drilled into us—on billboards, in documentaries, through the lips of politicians and influencers—until it feels like our job is to rescue something fragile, something teetering on the edge.

But the Earth isn't fragile. It's resilient, powerful, and constantly regenerating. It's been spinning through floods, fires, and ice ages long before we showed up, and it'll keep spinning long after we're gone. The question isn't whether the planet will survive—it will. The question is whether we, as individuals, families, and communities, will step into the roles we were meant to play within it.

We're not all called to be regenerative farmers, though I'll admit I sometimes wish everyone could feel the dirt under their nails the way I do now. We're not all meant to leave our lives behind, abandon the cities, and go live off-grid in the wilderness with nothing but a tent and a prayer. That's not the point. But we are all here, at this specific moment in time, for a reason. Modern life pulls us away from that reason—distracting us with the endless chase for money, status, fleeting recognition, the next promotion, the next shiny thing. It's a treadmill we can't step off, a noise so loud it drowns out the quiet voice inside. Yet, in those rare moments when we do step away—when the phone's off, the screens are dark,

I feel happier (and, I think, even look happier!) as a cattle rancher than I ever did as a vegan chef.

and the world hushes—we sense it. A stirring. A knowing. Something calling us back to what's real.

For some, that calling might be to tend the soil, to grow food, to provide nourishment for others. I felt that pull when I left Sage Vegan Bistro behind and sank my hands into the rocky earth of Fillmore. For others, it might be to nurture something else—a marriage through the hard seasons, a family through the chaos of raising kids, a community through the small, unglamorous acts of showing up. It could be as simple as growing herbs on a windowsill in a cramped apartment, watching those green shoots defy the concrete jungle outside.

These acts matter. They're not trivial. They tether us back to what has always sustained human life: food, family, faith, and community. They're threads in a tapestry we've let fray, but they're still there, waiting to be picked up.

The Stirring I Couldn't Ignore

I didn't always hear that call. Back in LA, running my restaurant, I was too busy checking boxes—sustainable sourcing, organic ingredients, reusable straws—to notice the emptiness gnawing at me. I thought I was doing my part, feeding people kale salads and oat-milk lattes, preaching a gospel of compassion that felt righteous at the time. But there was a moment, standing in my driveway beside those pomegranate trees I'd planted, when my brother Ryland called, his voice buzzing with excitement about soil and regeneration. That was the beginning.

It wasn't a thunderbolt. It was quieter than that—a whisper that grew louder the more I listened. When we finally bought that farm in Fillmore—twenty acres of rocky soil and unrealized dreams—it wasn't a leap of faith so much as a step I couldn't not take. The land was waiting, and so was I.

If you feel that stirring inside you—to plant a garden, to raise animals, to learn the rhythms of the land—don't ignore it. Follow it. If something awakens in you at the thought of producing local food, reclaiming lost knowledge, becoming a provider in a world of consumers—listen to that voice. It's not loud. It doesn't shout over the noise of your inbox or the news cycle. It's a quiet nudge, a pull toward something ancient and true.

The world needs more farmers, yes—people who'll dig into the soil and coax life from it, who'll resist the fragility of centralized systems that collapse when the trucks stop running. But even if your path doesn't lead you to a farm, you're still being called to tend to something. That's the universal thread—we're all stewards of something, whether we recognize it or not.

Tending the Literal and the Metaphorical

For me, that tending became literal—twenty acres of avocado trees, sheep grazing under the branches, compost piles humming with life. The first years were brutal—grief over the loss of my best friend Mimi, still running restaurants every single day, and taking on the new and not-easy job of farming. The hours never ended. The work never ended. I stood by Sespe Creek one afternoon, watching the water ripple past, and realized I was tending more than just soil. I was tending my family—my husband hauling rocks beside me, my children running through the orchard. I was tending a community—neighbors like Ernie King, who started as a skeptic but became family, sharing holidays and stories. I was tending myself—finding a faith I'd lost in the city, a trust in something bigger than my own plans. It wasn't just about growing avocados; it was about growing a life.

My family—with Ryland's family, my mother Jeanne, and her twin sister, Virginia. We belong here.

But that tending doesn't have to be a physical plot of land for everyone. It can be the sacred space of your own mind—cultivating thoughts that nourish instead of tear down. It can be a marriage, weathered by time but strengthened through the daily work of showing up. It can be a family, messy and loud, but held together by love and presence. It can be a community—neighbors you actually know, not just faces you pass on the street.

We weren't made to be passive consumers, endlessly dependent on systems that don't nourish us—grocery stores with food shipped from thousands of miles away, governments that dictate our choices, screens that keep us isolated. We were made to cultivate, to build, to care for, to create. That's in our bones, even if we've forgotten it.

We've drifted so far from that purpose. Look around—our food comes in plastic wrappers, shipped from factories we'll never see. Our families are fractured, stretched across states or screens, parents too busy to be present. Our communities are hollowed out, replaced by online echo chambers where we shout past each other. Our faith—whether in God, nature, or each other—has been swapped for fear, a nagging dread that we're too late, too broken, too small to matter. We've let ourselves be convinced that reconnection is optional, that we can outsource our lives to corporations and governments and still thrive. We can't.

I saw this disconnection up close during the COVID years. I watched my friends lose businesses and even their homes—and then I lost my businesses, one by one. I watched friends lose their hope—all because we'd built lives dependent on systems that could collapse overnight. Meanwhile, on the farm, we kept going. The soil didn't care about mandates. The sheep still needed feeding. The trees still bore fruit. We weren't invincible, but we were resilient. We had food we'd grown, neighbors we trusted, a rhythm that didn't break. That's what reconnection gives you: not immunity, but roots.

The cost of staying disconnected is higher than we admit. Our kids grow up on processed junk, their bodies and minds weaker for it—obesity, anxiety, autoimmune disorders climbing every year. Our marriages erode under the weight of distraction, our communities dissolve into anonymity. We've traded faith for control, thinking that technology—more apps, more gadgets, more rules—will save us. It won't.

Control is an illusion; the flood that hit our farm, with eighteen inches of water raging through, taught me that. My husband and the men waded in, saving what they could, while I made stew for when they stumbled back, soaked and exhausted. Nature doesn't bend to our will—it asks us to work with it.

Practical Steps Back

So, how do we reconnect? It's not about grand gestures—it's about small, deliberate steps.

Start with food. Grow something—anything—a tomato plant on your balcony, a patch of lettuce in your yard. If you can't grow it, buy it from someone who does—a farmer at a market, not a faceless chain. Cook it yourself, not from a box. Taste the difference. Feel the tether to the land snap back into place. I started with those pomegranate trees in my driveway, a modest food forest, and it changed everything. In just a few years, I was able to start harvesting $500,000 worth of produce off of only twenty acres—not because I'm a genius, but because I listened to the soil.

Tend your family next. Be present—put the phone down, eat together, talk. Hold your children; hug them. Sleep close to them; raise them close like mammals—not apart in cribs or schedules dictated by experts who don't know your child's cry. It's messy and loud, but it's real. That closeness builds something unbreakable—a bond that processed food and screen time can't touch.

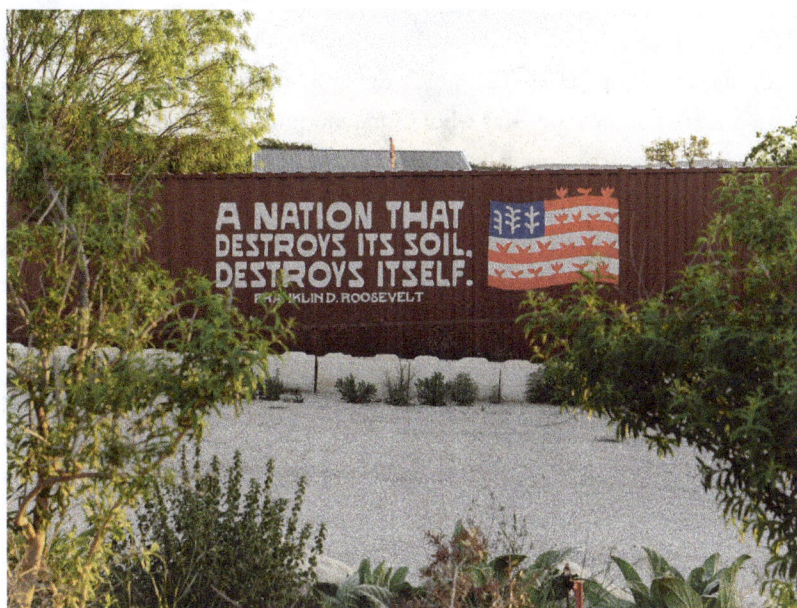

In our family, love for this country isn't optional. It's a core value.

Build community. Know your neighbors—not just their names, but their stories. When we moved to Fillmore, I braced for judgment—Trump signs and American flags lining Grand Avenue made me wonder what I'd walked into. But Ernie King, with his gruff advice and unexpected kindness, showed me I'd misjudged. Levi became a friend despite our politics, bonding over homeschooling and homesteading. The entire Stealy clan became like family; my neighbor Wendy and all the other women in my Azure Standard drop group, too. These weren't my "tribe" on paper, but they became my people. Start small—share a meal, lend a tool, show up when it counts.

And turn to faith, not fear. For me, it's God—felt most deeply with my hands in the soil, closer than any church pew. For you, it might be nature, the divine in the cycles of life, or trust in each other. Whatever it is, let it guide you. Fear paralyzes; faith moves. When I lost that first baby and faced marrying a man I barely knew, faith carried me through—not a hallucination, but a promise I made to something bigger. It's why we stayed, why we built a family, why I'm still here.

The Farm as a Mirror

The farm taught me these steps, but it's more than that—it's a mirror.

Every day, I see what reconnection looks like. The sheep graze, fertilizing the soil; the trees drop leaves that mulch the ground; the compost turns death into life. It's not separate—it's whole. One morning, I watched my daughter, seven years old, chase a black lamb across the field, laughing as it bleated back. She doesn't know disconnection yet—she's in it, part of it, alive to it. That's what we're aiming for: a life where we're not spectators but participants.

But it's hard. The farm's relentless—365 days a year, no weekends, no breaks. Animals need feeding, crops need tending, floods don't wait. I've stood in the rain, eight months pregnant, shouting directions as my husband wrestled a tractor to save our propane tank. I've cried over dead sheep, cursed dying trees, and wondered why I left the city. Reconnection isn't comfortable—it's calluses on your hands, sweat in your eyes, failure you can't hide from. Yet it's also abundance—avocados heavy on the branch, kids growing strong on raw milk, neighbors sharing a table. It's worth it.

We don't all need to farm, but we all need to steward. The divine—whether you call it God, nature, or something else—doesn't ask us to have all the answers. It asks us to listen, to follow, to take the next step. When I started, I didn't know soil organic matter from a hole in the ground—ours was less than 1 percent when we arrived. A few years later, with grazing and wood chips, some of it is 20 percent. I didn't fix the planet; I tended my patch. That's enough.

So I ask you: What's calling you back? It might be a literal plot—carrots in a raised bed, chickens in your backyard. It might be your kids, your partner, your street. Whatever it is, step toward it. Build community—share a harvest, a story, a hand. Grow food—even a sprig of basil counts. Raise healthy families—messy, loud, real ones. Allow yourself to be led—not by fear or control, but by faith, by stewardship, by the quiet voice that knows.

The way forward isn't through more consumption—another Tesla, another app—or more technology promising to outsmart nature. It's through reconnection, through becoming the best possible cell in the body of this Earth.

We're not here to save the world; we're here to belong to it. Find your way back. The time is now.

Stay Connected!

Thank you for walking with me through *Debunked by Nature*. If this book has stirred something in you, I invite you to take the next step with the *Debunked by Nature Companion Workbook*, available for download at sovereigntyranch.com/product-page/workbook or bookstore.acresusa.com.

The workbook is designed to help you reflect, go deeper, and apply these principles to your own life in a meaningful and practical way.

To stay connected and to see what's unfolding at Sovereignty Ranch—from regenerative farming and hospitality to future projects, events, and resources—you can always visit sovereigntyranch.com.

And follow along here:

- Instagram: @talktoMollie
- YouTube: @talktoMollie
- Twitter / X: @talktoMollie

Thank you for reading, and thanks for being part of this conversation about nature, life, and truth.

With gratitude,

Mollie Engelhart

sovereigntyranch.com/product-page/workbook

Other Titles from Acres U.S.A. Books:

Restoration Agriculture
By Mark Shepard

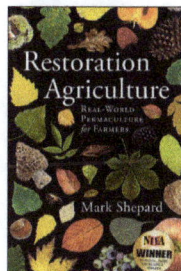

Around the globe, most people get their calories from "annual" agriculture—plants that grow fast for one season, produce lots of seeds, then die. Every single human society that has relied on annual crops for staple foods has collapsed. *Restoration Agriculture* explains how we can have all of the benefits of natural, perennial ecosystems and create agricultural systems that imitate nature in form and function while still providing for our food, building, fuel and many other needs—in your own backyard, farm, or ranch. Based on real-world practices, this book presents an alternative to the agriculture system of eradication and offers exciting hope for our future.

Growing Life
By André Leu

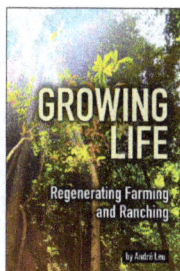

As modern farming and ranching evolve away from mass consolidation and industrialization, a new strategy is rapidly emerging: regenerative agriculture. These new systems being implemented across the globe require a shift in the mindset of the land manager and operator—away from being primarily reliant on external inputs, such as fertilizers and pesticides, and toward dependence on knowledge, measurement, and management. André Leu invites everyone to start moving in a positive, regenerative direction, where our actions support growing abundant life.

The Farm as Ecosystem
By Jerry Brunetti

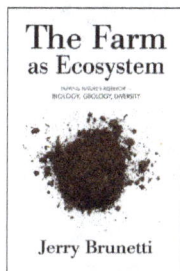

In *The Farm as Ecosystem*, natural product formulator and farm consultant Jerry Brunetti brings together a wealth of education and uncanny observations on the interconnected dynamics of the farm—geology, biology, and diversity of life. Learn to look at—and manage—your farm very differently by gaining a deeper understanding of the complementary roles of all facets of your operation. The author guides the reader on a journey through the modern farm as an ecosystem, providing intimate anecdotes and comprehensive details that appreciate all dimensions of the farm. Brunetti's work is invaluable to the contemporary farmer and to those seeking an original appraisal of farming and its future.

bookstore.acresusa.com

Visit acresusa.com to learn about our magazine, events, books, and other educational opportunities.

About Acres U.S.A.

Founded in 1971 by Charles Walters, Acres U.S.A. emerged from the need to promote ecological farming practices in a time when industrial agriculture was heavily reliant on synthetic fertilizers and pesticides. Insoired by figures like Rachel Carson and Dr. William Albrecht, Walters used the magazine, and later books and conferences, to advocate for sustainable agriculture that prioritized soil health and natural processes. Acres U.S.A. provided a platform for these ideas and helped to popularize alternative methods like cover cropping and integrated livestock management.

Though the agricultual landscape still relies heavily on conventional methods, Acres U.S.A. has been instrumental in the growing movement towards regenerative agriculture. By disseminationg knowledge and supporting eco-conscious farmers, the company continues to champion sustainable practices through its publications, conferences, and online resources, contributing to a shift towards a more environmentally sound approach to farming.

Find Out More About Acres U.S.A.

Subscribe To The Online Magazine'
https://members.acresusa.com/

Attend The Eco-Ag Event
https://acresusa.com/events/

Visit The Acrea U.S.A. Bookstore
https://bookstore.acresusa.com/

Join The Free Newsletter
https://mailchi.mp/acresusa/newsletters

www.ingramcontent.com/pod-product-compliance
Lightning Source LLC
Chambersburg PA
CBHW072255270326
41930CB00010B/2384